Assured Survival

**RELATED BOOKS
BY BEN BOVA**

Fiction

*Millennium
Colony
Kinsman*

Nonfiction

*The Fourth State of Matter
The New Astronomies
Closeup: New Worlds
The High Road*

ASSURED SURVIVAL

Putting the Star Wars Defense in Perspective

Ben Bova

Houghton Mifflin Company   Boston   1984

To my father

Copyright © 1984 by Ben Bova

All rights reserved. No part of this work may be reproduced or transmitted in any form or by any means, electronic or mechanical, including photocopying and recording, or by any information storage or retrieval system, except as may be expressly permitted by the 1976 Copyright Act or in writing from the publisher. Requests for permission should be addressed in writing to Houghton Mifflin Company, 2 Park Street, Boston, Massachusetts 02108.

Library of Congress Cataloging in Publication Data

Bova, Ben, date
Assured survival.
Bibliography: p.
1. Ballistic missile defenses — United States.
2. Atomic warfare. 3. Space weapons. 4. United States — Military policy. I. Title.
UG743.B68   1984   358'.17'0973   84-10491
ISBN 0-395-36405-1

Printed in the United States of America

Q 10 9 8 7 6 5 4 3 2 1

The author is grateful for permission to quote from the following sources:
The Wizards of Armageddon: Strategists of the Nuclear Age by Fred Kaplan (New York: Simon & Schuster, 1983); "War in Space: Good!" by Stewart Brand (*New Scientist*, 1981); and "The Danger of Thermonuclear War" by Dr. Andrei Sakharov (*Foreign Affairs*, Summer 1983).

Acknowledgments

A VERY WISE AUTHOR, when asked how long it had taken him to write his latest book, replied, "All my life." Similarly, many of my friends and associates have been helping me to write *this* book for many, many years — even though neither they nor I knew it at the time. Then there are those who graciously gave me some of their valuable time specifically to be interviewed for this book. I thank them all and list below those who contributed the most to making *Assured Survival* the book that it is. The factual content and personal views that they have shared with me have been invaluable in creating this work. On the other hand, if I have erred in reporting their views or unwittingly misrepresented the facts, the fault is mine, not theirs. Memory being less than perfect, I have undoubtedly left out many people whose help was just as important as those listed below; I apologize in advance for the oversight and hope that they will forgive me.

I extend my special thanks, therefore, to Gregory Benford, Robert S. Cooper, Frederick C. Durant III, Edward T. Gerry, Daniel O. Graham, Jerry Grey, Arthur R. Kantrowitz, George A. Keyworth II, Michael G. Michaud, James Muncy, James E. Oberg, Jerry E. Pournelle, Victor M. Reis, Stanley Rittenberg, Stanley Rosen, Jeff Rovin, and Kosta Tsipis. To list all their accumulated degrees, titles, and achievements would take many, many pages; suffice it to say that they are all tremendously accomplished men, and I deeply appreciate their help and generosity.

I must also single out my editor, Austin Olney, whose encouragement, advice, and thoughtful criticism have helped to shape and strengthen this book. At a time when book publishing has become almost as impersonal as a supermarket, his warm friendship and guidance are rare and treasured.

More than a little credit should go to Ernest Gardner, architect, and Charles Gregory, general contractor, who between them made certain that I would be up early every morning and at my desk, writing.

Finally, my deepest appreciation goes to my agent, most perceptive critic, strongest supporter, wife, and lover, Barbara. Without her, this book would never have been started, and my personal "assured survival" would be nowhere as important to me as it is now.

Contents

Acknowledgments   v

Introduction: Freedom from Fear   1

PART I: THE NIGHTMARE AND THE DREAM

1   Scenario One: The Ultimate Fear   5
2   Replay   13
3   The Real World   19

PART II: THE ROAD TO NOW

4   Countdown to Armageddon   39
5   Why Fortify Heaven?   89
6   Naked to Mine Enemies   99
7   "Star Wars" vs. "MAD"   115
8   The Speech of 23 March 1983   129
9   The Pros and Cons   149
10 Smart Weapons   176

PART III: DIALOGUES

11 Scenario Two: The European Face-Off   195
12 No. 14 Borodin Avenue, Kutusow Prospekt   206
13 The Russian Decision   220
14 19, St. Mark's Crescent, Regent's Park   238
15 The Soldier's Dilemma   251

PART IV: THE PATH TO PEACE

16 Scenario Three: A Good Strong Roof   271
17 Where We Stand   275
18 Scenario Four: The Great Divide   297
19 Warfare Suppression   311

Epilogue: The Athenian Ideal   323

Appendix: Space Weapons   329

Bibliography   342

Introduction: Freedom from Fear

Out of this nettle, danger, we pluck this flower, safety.
— William Shakespeare

The message of this book is this: we have the means almost within our grasp to put an end to war. Not merely nuclear war, but conventional wars as well. The forty-year nightmare that we have all undergone can be ended. The nuclear holocaust that we all dread can be averted. Permanently.

Since Hiroshima, the technology of armaments has been creating weapons of constantly more dreadful power, weapons of attack that can destroy all human life. Now, almost forty years after the first atomic bomb sent its mushroom cloud rearing into the New Mexico sky, the technology of armaments is beginning to create weapons of defense, weapons that may be able to shield us against the looming threat of nuclear Armageddon and even against the "conventional" horrors of non-nuclear warfare.

The tools for *warfare suppression* are being developed today, in laboratories and testing grounds on both sides of the Iron Curtain and even in orbital space. Strangely, very few of the men and women working to create these tools think of them as means for ending war; they believe they are developing better weapons for fighting future wars. And stranger still, some men who have spent their lives furthering the development and use of fearful offensive weapons are bitterly opposed to developing the tools for defense.

These defensive tools include high-power lasers and other so-called "directed-energy weapons"; new computers that are smaller, faster, and more rugged than any previously developed; drone aircraft and missiles that are "smart" enough to find and destroy even the most elusive and heavily defended targets on land, sea, or in the air. Such defensive weapons may make it impossible for an aggressor to launch an attack, whether it be a nuclear missile strike or a massing of tanks to cross an international border.

But the tools alone are not enough. Weapons, even defensive weapons, cannot put an end to war by themselves. We must match these new developments in technology with new developments in our political outlook, with new ideas that are as bold and sophisticated as the tools now being created.

If we use the knowledge we already have and match it with wisdom, courage, and foresight, we can create a world that is truly free from the fear of war, free from the nightmare of nuclear extinction that haunts our dreams.

PART I

The Nightmare and the Dream

To employ a mathematical analogy, we can say that although the risk of extinction may be fractional, the stake is, humanly speaking, infinite.

— Jonathan Schell

1

Scenario One: The Ultimate Fear

IT WAS the death of Ayatollah Khomeini that led to the end of the world.

Deprived of their charismatic symbol of the Islamic Revolution, Iran's Revolutionary Council found itself beset on all sides. The national economy, which had slowly been improving despite the downward trend in world oil prices, faltered badly. The Communists began demonstrating openly in the streets once again. The interminable war with Iraq had brought nothing but mounting casualties and increasing damage to Iran's oil-production facilities. The people were restless and reaching the end of their patience with shortages, political uncertainty, and the grinding attrition of the war.

In Lebanon, meanwhile, Israel's policy teetered between a further withdrawal of its occupation troops back to the Lebanon-Israel border and a full-scale war with Syria. For more than two years the Israelis had tried to find a way out of Lebanon. Crushing the Palestine Liberation Organization (PLO) as a military force had merely brought Israel into a direct confrontation with Syria. The so-called peacekeeping forces sent to Beirut by the United States, Britain, France, and Italy had become little more than targets for terrorist bombs and snipers. Washington withdrew the American Marines from Lebanon while trying to avoid the appearance of capitulating to Soviet-backed Syria. And the other troops were pulled out, too. The assassination of Yasir Arafat, after he was driven out of Tripoli, completed the Syrian domination of the PLO. Israeli leaders knew that a withdrawal of their troops to their own frontier would mean only that the war, when it began, would start on Israel's soil.

Elections in Israel had ushered in a new government, but not new policy decisions. It was obvious that the constant war of nerves and the attrition caused by guerrilla and terrorist attacks on Israeli troops in Lebanon could not be allowed to continue. World opinion was turning slowly but inexorably against a continued Israeli occupation of Lebanese territory, and Washington was constantly putting more pressure on the Israelis to withdraw.

It was never made clear just who took the first step to bring about the cease-fire between Iran and Iraq. On the scant evidence that was available before the holocaust destroyed everything, it appeared that the Syrians approached both governments in the name of Pan-Islamism, seeking their support for "the final attack" on Israel. Certainly the accord between Syria and Qaddafi's Libya, widely trumpeted in the world's media, started from a Syrian initiative.

Israel's intelligence service quickly perceived that Iran and Iraq had not merely agreed to a cease-fire, but had pledged their forces for the war against Israel. But no one in Jerusalem expected the war to begin in the Sahara, with an attack by Libya against Egypt's western frontier.

Egypt, the sole Islamic nation to have come to peace terms with Israel, uneasily swung its main military forces

Scenario One: The Ultimate Fear

away from the Sinai Peninsula to meet the invasion of Russian-built T-72 tanks churning up the desert sands by the hundreds.

Jordan's wily King Hussein escaped an assassination attempt when a bomb wired to his limousine exploded prematurely, but Hussein realized the true meaning of the attempt on his life. He publicly condemned Israel's continued occupation of southern Lebanon and secretly guaranteed Syria that Jordan would allow Syrian forces to cross its territory in a flanking attack on Israel.

Seeing itself surrounded by hostile forces, all of them armed by the Soviet Union, the Israeli government, while publicly warning of the grave consequences of a new Middle East war, secretly sent a personal representative of the new Prime Minister to Damascus to discuss possible settlement terms with the ailing President al-Assad. The discussion lasted less than one day. When Egypt's President Mubarak was abruptly overthrown by a coup of his own army officers, the Prime Minister's representative hurried back to Jerusalem for further instructions.

The new Egyptian government renounced the Camp David Accord with Israel, and Egyptian and Libyan soldiers embraced each other in the western desert. The ring of steel around Israel was now complete.

While the Knesset was locked in bitter debate over Israel's choice of war or negotiation, fighting broke out in the Golan Heights. Who struck the first blow, it is impossible to say. The morning began with artillery bombardments. By noon, Syrian tanks and armored infantry were attacking entrenched Israeli positions on the heights while the sky above became an intricate lacework of jet contrails, blackened here and there by the burning plumes of warplanes making their final plunge to earth.

Within less than twelve hours, Israeli troops in southern Lebanon were being pushed back by the combined Syrian-

Iraqi-Iranian forces. Pressed hard, losing ground, and — more important to a nation of only four million people — suffering high casualties, Jerusalem warned that it would use nuclear weapons against its enemies if and when they crossed the borders of Israel.

Damascus countered immediately with a threat to "meet nuclear force with nuclear force." The rest of the world became frantic with dread. The hot line between Washington and Moscow blazed all through the night with urgent messages. Moscow insisted that the United States stop Israel from "taking the first step to a worldwide holocaust." The White House, in a sternly worded message, warned Israel that if it used even a single nuclear weapon, all American aid to Israel would cease immediately, and Israel would be considered "an outlaw nation" by the entire world.

For half of the second day of the war, Israel withheld its nuclear hand. But when Syrian tanks and artillery finally battered the last Israeli defenders from the Golan Heights, and the new Egyptian government began moving its armored columns from the Libyan border to the Sinai, the Prime Minister told his people by radio and television from Jerusalem that he had no choice: the very existence of Israel was at stake.

Two small, low-yield nuclear weapons, carried by drone aircraft, obliterated the Syrian forces atop the Golan Heights. The twin mushroom clouds rose into the sky like symbols of Armageddon — which they truly were.

As the sun was setting over those ravaged heights, the city of Haifa was blasted off the face of the earth by a nuclear explosion estimated to be in the 150-kiloton range — almost ten times more powerful than the bomb that destroyed Hiroshima. American surveillance satellites tracked a missile launch from northwestern Iran, between

Scenario One: The Ultimate Fear

the city of Tabriz and the Russian border. It had taken only five minutes for the missile to reach Haifa.

More than fifty thousand were killed immediately in the blast and fire that raged uncontrolled through Haifa. Twice that number would eventually die of burns, injuries, or radiation poisoning.

The American government publicly blamed the Soviet Union for providing the nuclear weapon, the missile, and the crew that launched it from Iranian soil. But the United States refrained from any kind of retaliation against the USSR, calling instead for an international tribunal to investigate the matter and "punish appropriately" the nations responsible for the devastation of Haifa.

Seeing themselves alone and facing total defeat, in the dark hours just before dawn the Israelis annihilated Damascus with a 1-megaton hydrogen bomb, delivered by a low-flying jet attack plane that was either destroyed in the explosion or shot down by Syrian defenses on its way back to its base.

As the first rays of sunlight touched the gilded turrets of the Kremlin, Moscow announced to the world that it would provide "all aid, of every kind," to its Arab friends. Jerusalem responded by declaring that the next nuclear weapon used against them would result in a retaliation "against the source of that weapon."

That was the first forty-eight hours of the war.

To forestall a combined Egyptian-Libyan attack through the Sinai, the desperate Israelis launched a barrage of six nuclear missiles against Egyptian positions in key mountain passes on the peninsula, both to prevent an attack on a second front and to make the critical passes through the region too radioactive to be used for several weeks.

Tel Aviv was the next victim of nuclear war, wiped out by a missile launched from a submarine operating in the western Mediterranean Sea. Israel retaliated by sending a suicidal air strike against Sevastopol and Odessa, the only major Russian cities that Israeli warplanes had any hope of reaching. Soviet air defenses wiped out the six-plane group heading for Odessa while the planes were still over the Black Sea. But two planes of the nine sent against Sevastopol got through, and the city was virtually destroyed by a pair of 100-kiloton bombs. Nearly seventy thousand Russians were killed immediately.

Moscow issued an ultimatum: if Israel did not surrender to Syria within six hours, Soviet missiles would destroy the nation. The Russians simultaneously warned the United States and Western Europe not to intervene.

The pressures and counterpressures in Washington were enormous. Faced with the possibility of Israel's destruction and total Soviet domination of the oil-rich Middle East, the White House sent a vaguely worded message to Moscow, warning that nuclear escalation "can only lead to a wider circle of destruction."

The Russians responded to Washington's message by launching a pre-emptive first-wave nuclear attack at the United States two hours *before* the deadline for the Israeli surrender.

Seeing more than fifteen hundred nuclear warheads on their way to targets in America, the President collapsed. It fell to the Vice-President to order a full retaliatory strike on the Soviet Union. It was his last official act — and the last act of his life.

The first Russian strike was aimed at "decapitating" the American government and destroying as much of the American strategic forces as possible. Washington was the only major city to be directly attacked. The other Soviet missiles were aimed at American missile bases, command centers, airfields, and submarine bases.

The American counterattack obliterated the Soviet

Scenario One: The Ultimate Fear

Union as a modern nation. But not before the Russian second-wave attack destroyed every major city in the United States and devastated the whole continent with lethal levels of radiation.

Europe would have been spared if NATO commanders (presumably the Americans among them) had not launched their nuclear missiles at prearranged targets in Eastern Europe. Medium-range Soviet missiles rained out of the sky for half an hour, leaving Rome, Paris, Copenhagen, London — the entire continent of Europe — little more than a smoking radioactive wasteland.

Nearly a billion people were killed within little more than an hour by blast and fire. Billions more, particularly in China and Japan, were pelted with lethal radioactive fallout. A huge pall of smoke and ash blotted out the sun. Radioactive debris from the ground bursts that had smashed fields of missile silos joined with the black smoke from burning oil refineries and the vast forest fires that swept Siberia and North America. For weeks there was no sunlight in the northern half of the globe. "Nuclear Winter" began, and crops that had not been burned or blasted away wilted and died from lack of light and the deepening, numbing cold. The high-altitude nuclear explosions that had blanked out defensive radars and knocked out all sorts of electrical equipment with their continent-spanning electromagnetic pulses had also destroyed much of the earth's ozone layer; unfiltered solar ultraviolet radiation burned crops, animals, and people in the regions of the globe not covered by the clouds of Nuclear Winter.

Temperatures plummeted everywhere. Gray radioactive snow fell in the ruins of Miami and Honolulu. The black clouds of death spread southward, covering the earth with the hand of death. Fields of grain withered and froze. Herds of animals died of starvation, of cold, of radiation poisoning. The last pitiful survivors of the human race slowly succumbed as they huddled in terror in caves or mines deep underground.

The horrifying prediction of Jonathan Schell in his book, *The Fate of the Earth,* came true:

> Bearing in mind that the possible consequences of the detonations of thousands of megatons of nuclear explosives include the blinding of insects, birds, and beasts all over the world; the extinction of many ocean species, among them some at the base of the food chain; the temporary or permanent alteration of the climate of the globe, with the . . . chance of "dramatic" and "major" alterations in the structure of the atmosphere; the pollution of the whole ecosphere with oxides of nitrogen; the incapacitation in ten minutes of unprotected people . . . the blinding of people who go out into the sunlight; a significant decrease in photosynthesis in plants around the world; the scalding and killing of many crops; the increase in rates of cancer and mutation around the world, but especially in the targeted zones, and the attendant risk of global epidemics; the possible poisoning of all vertebrates . . . as a result of increased ultraviolet light [from destruction of the ozone layer in the high atmosphere]; and the outright slaughter on all targeted continents of most human beings and other living things by the initial nuclear radiation, the fireballs, the thermal pulses, the blast waves, the mass fires, and the fallout from the explosions; and, considering that these consequences will all interact with one another in unguessable ways and, furthermore, are in all likelihood an incomplete list . . . one must conclude that a full-scale nuclear holocaust could lead to the extinction of mankind.

2
Replay

THERE ARE a thousand ways to start the end of the world. The scenario in Chapter 1 is merely one of them, perhaps a little more likely than some, a little less realistic than others.

Nuclear holocaust can come at any time, for almost any reason: a guerrilla war in Central America, a civil uprising in Poland, a moment of madness in the White House or the Kremlin, an accidental malfunction of early-warning radars or their complex computerized defense networks.

Today, at this moment, if the nuclear-armed missiles are launched from their silos and submarines — for whatever reason — the end of civilization is only thirty minutes away. There is no possibility of recalling those missiles once they are launched. There is no way to stop them or to defend against them.

Today.

But what about tomorrow?

Let us suppose, as a thought experiment, that there is a defense against ballistic missiles. Further, let us replay the Armageddon scenario of Chapter 1 with the addition not only of such an antiballistic missile (ABM) defense, but with a whole panoply of *warfare suppression* technology in place — together with an international peacekeeping force that has the political power to use such technology.

Orbiting a few hundred miles above the earth's surface, a half-dozen surveillance satellites scan land and sea with sensitive cameras and other sensors. Each satellite moves in a polar orbit that allows it to see the entire earth, from North Pole to South, every twelve hours as our planet revolves majestically beneath the tiny man-made moons.

The satellites are part of the network of surveillance systems, communications links, and automated weaponry of the newly formed International Peacekeeping Force (IPF). They have been in orbit for less than a year, and in that time they have kept watch over the sporadic artillery duels and small-scale raids that have simmered along the battlefront between Iran and Iraq.

Now the satellites report back to their controllers on the ground that significant troop movements are being made in both Iraq and Iran. The analysts are surprised, at first, to see that the troops are not massing to attack one another, but are forming motorized columns and heading west, toward Syria.

As the IPF goes on alert status and sends queries to Tehran and Baghdad, the surveillance satellites spot armor concentrations building up on Libya's side of its border with Egypt. A warning is issued immediately, but no reply comes from Tripoli. IPF headquarters reports the situation to the United Nations, the World Court, and to the world's media.

Israel, knowing full well what these troop movements imply, announces that it will comply fully with the IPF efforts to prevent a new Middle East war, but that it is mobilizing its own forces, nonetheless. The feeling in Jerusalem — and most of the other capitals of the world's nations — is that the fledgling IPF may be able to detect troop movements and issue warnings, but it cannot prevent nations from going to war.

What none of the politicians realizes is that, although the peacekeeping force may not be able to prevent a war from starting, it has the potential power to *end* a war very quickly.

When Libyan tanks cross the frontier into Egypt, they are met with swarms of tiny drone missiles, launched by the hundreds from IPF ships well out to sea. Like a plague of locusts, these missiles seek out the lumbering tanks and destroy them, smashing through their thickest armor to explode inside the tanks, killing their crews and setting off their ammunition and fuel. Within an hour, the western desert is dotted with the smoky funeral pyres of hundreds of tanks. Those which are not destroyed in the first wave of the missile onslaught retreat back to Libyan soil.

As the Libyan invasion fizzles, the Egyptian army's planned coup against President Mubarak also runs aground. The clique of army officers who were plotting against their President are rounded up, and the main strength of the army remains loyal to Mubarak.

The shelling that opens hostilities along the Golan Heights is quickly suppressed by a rain of similar missiles, launched from IPF planes patrolling high above the waters of the western Mediterranean. Both Syrian and Israeli artillery pieces are shattered by the tiny, hypervelocity missiles. Casualties among the gun crews are heavy at first, but when it becomes apparent that the IPF missiles are methodically destroying every artillery piece in the region, the men abandon their guns and save their own lives.

For several days the situation remains tense. The armies of Syria, Iran, and Iraq are coiled like loaded springs, waiting for the order to attack. Israel is at full mobilization, its guns leveled at its neighbors to the north and east; but with a wary eye it still glances uneasily back toward the Sinai, even though all seems quiet on that front. To outside observers, it seems as if they all *want* to fight, but no one dares to make the first move, for fear of drawing the punishment of the IPF's clouds of cheap, deadly, "smart" missiles.

Then the nuclear missile is launched from northwestern Iran. Moscow later claims that Islamic fanatics overpowered the Russian missile crew and forced them to launch it, against the Kremlin's strictest orders. The world will probably never know what actually happened, because the entire launching crew was either murdered in Iran or shot after a summary military trial in the Soviet Union after returning home.

In any event, the missile is launched. It is an SS-20 intermediate-range ballistic missile, carrying only a single 150-kiloton warhead, aimed at the port city of Haifa. There are some cynics, particularly in Washington, Bonn, and Paris, who believe that the Russians deliberately launched the missile to see what the IPF was capable of doing. Others accept the explanation that Moslem fanatics were responsible for the launch; that they hoped to force the final destruction of Israel, despite the stalemate brought about by the IPF.

Although the motivations and responsibility for the missile launch remain cloudy, the IPF's reaction is quite clear. Several satellites immediately detect the launch and relay the information automatically both to IPF headquarters in Geneva and to the laser-armed ABM satellites that orbit the globe.

Much faster than any human could decide or move, the IPF's computerized command-and-control system assigns one particular laser satellite to destroy the missile, and a second satellite to stand by as a backup to the first.

Within a few thousandths of a second, the computers aboard the laser satellite assess the situation. The satellite's tracking telescopes lock onto the missile while its rocket engines are still bellowing flame. The laser's thirty-foot-wide copper mirror swivels silently in its mounting, following the missile's unswerving flight. Deep within the bowels of the massive satellite, its electric power generator vibrates to life, energizing the laser itself.

A beam of invisible energy lances across more than a thousand miles of empty space, from the satellite's laser mirror to the aluminum skin of the still-rising missile. The beam is pulsed, like the staccato of bullets spitting from a machine gun, except that each pulse of laser energy is incredibly smaller than a bullet, and the pulses are only microseconds apart.

The first pulse of laser energy to hit the missile scorches the paint off the aluminum and blackens the metal. The next few pulses smash through the aluminum altogether. The rocket engine's liquid propellants spill out, the engine falters, and suddenly a brilliant fireball explosion blossoms silently in the vacuum of space.

The missile's warhead, bearing the 150-kiloton nuclear weapon, was built to withstand the heat and pressure of re-entry into the atmosphere. The explosion of its carrier missile does not bother it much. It tumbles wildly back to earth and explodes at its preset altitude, two thousand feet above the desolate mountainous country where the borders of Iran, Iraq, and Turkey join. If any isolated shepherds or nomadic mountain people are killed by the blast, there is no report of the deaths. Radioactive fallout spreads in a wide oval for several hundred miles to the southeast, and there will certainly be deaths from radiation poisoning among innocent Iranian and Iraqi civilians.

But the IPF has demonstrated that it has the power to stop a nuclear-armed missile. Haifa survives. The war that might have led to nuclear Armageddon is stifled before it begins.

The world faces a curious situation now. Armed to the teeth, the nations seem unable to start a war. Even in the seething Middle East, where violence has been a way of life since biblical times, Israel and its neighbors face each other with no less enmity, *but they are unable to fight.* International rumor has it that there may even be a peace conference soon. No one expects much progress from it, but facing one another across a conference table is better than facing one another across a battle line, everyone seems to agree.

That scenario seems much more fanciful than the nightmare of Chapter 1. It seems much more plausible to expect nuclear war and the destruction of our world than to expect an International Peacekeeping Force that can, and will, prevent war. "Too good to be true" is the common human reaction to such a concept. People who accept the idea of nuclear war, who dread the looming imminence of worldwide holocaust, find it hard to believe that there are ways to prevent war from starting. "Too good to be true," they say. There is no such phrase in the language as "too *bad* to be true." We readily accept news of catastrophe; we hesitate to believe that catastrophes can be averted.

Yet the technology of our imaginary IPF is being brought into existence today. Drone aircraft, smart missiles, surveillance satellites, communications satellites, and even energy-beam weapons capable of destroying ballistic missiles in flight either are already available or are undergoing intensive development efforts.

As far as the technology is concerned, the tools for warfare suppression can be built. That is the easiest part of the task.

3

The Real World

"I BELIEVE we can counter the awesome Soviet missile threat with measures that are defensive," said President Reagan in his speech of 23 March 1983. Wouldn't it be wonderful, the President asked rhetorically, if we could "intercept and destroy strategic ballistic missiles before they reached our own soil or that of our allies."

Acknowledging that such defensive measures constitute "a formidable technical task, one that may not be accomplished before the end of this century," the President said, "Yet current technology has attained a level of sophistication where it is reasonable for us to begin this effort."

President Reagan's speech, and subsequent briefings given by White House aides to the media, brought forth a torrent of denunciation, applause, warnings, and comments by politicians, pundits, and thinkers of every stripe and caliber. Most of those who rushed into public print or on the airwaves had been caught by surprise: the idea of placing satellites in orbit armed with lasers or other weapons capable of destroying nuclear bomb–carrying ballistic missiles seemed totally new and fantastic to them.

The media quickly dubbed it a "Star Wars" concept, after the popular science-fantasy film.

At first I smiled at the appellation. In bygone years, any forward-looking concept involving new technology had been dubbed a "Buck Rogers scheme." At least the media had updated its sarcasm. Then I began to understand that the nickname was merely the result of a generation gap: many of the reporters had only vaguely heard of Buck Rogers, who was a hero of science fiction in the 1930s and forties. *Star Wars* was *their* generation's futuristic fantasy.

In the weeks following 23 March 1983, I found myself in a rather unusual position. Unlike most of the people making public comments on the President's proposal, I knew something about the technology of high-power lasers and their potential military uses. I had been the manager of marketing for the laboratory where the first high-power laser was invented. I had helped to arrange the first top secret briefing in the Pentagon on the subject. I was privy to the first studies sponsored by the Air Force on how such high-power lasers might be used against ballistic missiles. And I am also an author. Through those years and ever since, I have written books — both nonfiction works and futuristic novels — that examine the military and political effects that such weaponry could have.

I was in such an advantageous position because I was fortunate enough to spend a dozen years of my life in a unique place, the Avco Everett Research Laboratory. The laboratory, located in Everett, Massachusetts, a few miles north of Boston, was the creation of an extraordinary man, Arthur R. Kantrowitz. He founded Avco Everett in 1955 in an abandoned tire warehouse next to the Medford city dump and built it into one of the nation's top research organizations. I joined the lab in 1959, as a science writer, and then organized its marketing department.

The Real World

Avco Everett had been created specifically to solve the problem of missile re-entry. In the early 1950s American intelligence sources were startled to find that the Russians were testing very large, long-range rockets. It was clear that before the decade was out, the Soviets would possess rocket-driven missiles that could drop hydrogen bombs on the heartland of the United States. In deep secrecy, the Defense Department started a crash program to build similar rocket boosters. A new term sneaked into the language: ICBM, standing for intercontinental ballistic missile.

The Air Force was put in charge of the project and quickly realized that there were three crucial problems to be solved: propulsion, guidance, and re-entry. Rocket engines could be developed and tested on the ground. Electronic guidance systems could be built and tested aboard airplanes. But no one could figure out how to build and then test a re-entry vehicle, a detachable "nose cone" that could carry the missile's hydrogen bomb warhead through the blazing heat and shock of re-entry.

The re-entry vehicle sits atop the eight-story-tall missile as its rocket engines blast it several hundred miles up into space. In that airless, silent vacuum the re-entry vehicle (or R/V, in the aerospace engineers' jargon) is detached from the main body of the booster rocket and glides onward by itself. Not much taller than an average-size man, the R/V coasts along its ballistic path until the force of the earth's gravity and its own forward momentum make it dip back into the atmosphere. At this point it is traveling at well over ten thousand miles per hour. It hits the upper layers of the atmosphere like a rifle bullet smacking into a brick wall. Meteors falling out of deep space enter the atmosphere at similar speeds and are burned into dust long before they get near the ground. We call them falling stars. The re-entry vehicle must somehow endure that fiery passage, must survive the blazing heat generated by its hypersonic dive into the atmosphere, heat so intense that it turns the air around the R/V into glowing gases. The death-dealing machinery inside the R/V must not get so hot or so rattled by shock and vibration that it fails to detonate the hydrogen bomb at its preset altitude.

In 1955, no one except the Russians knew how to build such a re-entry vehicle. And, of course, no one knew how to test one, short of putting an experimental R/V on a full-size missile and flying it over its five-thousand-mile range to see what happens. Only the Russians had flown such a vehicle, and therefore only they had any understanding of how the atmosphere behaves when it is suddenly heated to temperatures that break down the air into incandescent, ionized gases.

Arthur Kantrowitz was a professor of engineering physics at Cornell University in 1955. He was known as something of a maverick, a scientist who told his students that he was not terribly interested in pure science, but preferred instead "impure science" — applied research, scientific investigations that had a specific goal in mind, rather than fundamental explorations into totally unknown areas of science.

Legend has it that Kantrowitz got into a conversation at a cocktail party on the Cornell campus with an Air Force officer and the president of Avco Corporation, Victor Emmanuel. The Air Force man worried aloud about the re-entry problem. Kantrowitz said airily, "Oh, I could solve that problem for you in six months." The officer said that the Air Force would be quite happy to fund six months of research if it led to a solution of the re-entry problem. Victor Emmanuel offered Kantrowitz a free hand to set up a laboratory wherever he wanted to and to hire a staff of his own choosing.

Thus was born Avco Everett Research Laboratory. Kantrowitz picked the warehouse in Everett because it was close to Harvard, MIT, and the other fine universities of the Boston area — and because of the region's good sailing. He was, and still is, happiest at the helm of a sleek sailboat. His staff consisted largely of his Cornell graduate students, a bright and aggressive crew of men young enough not to be daunted by the impossible. At Cornell, Kantrowitz and his students had been studying the behavior of various gases at very high temperatures, using devices called *shock tubes*. Very simple in their basic nature, shock tubes look like, and are, little more than lengths of pipe. Once, when I was conducting a visiting VIP through the laboratory and mentioned that the particular shock tube we were admiring looked rather like a sewer pipe, the technician working on that tube snarled, "Most expensive goddamned sewer pipe *you* ever saw!" Loyalty ran high at Avco Everett.

In six hectic months, Kantrowitz and his staff used those shock tubes to probe the behavior of shock-heated air and its constituent gases. They learned enough to be able to tell the Air Force how to design a re-entry vehicle that would work. The Air Force presented Avco Corporation with a check for $111 million to design and build the first nose cones. By late 1958 the first American ICBM flew some four thousand miles down the Atlantic Test Range, from Cape Canaveral to an impact site near Ascension Island. The re-entry vehicle worked as it was designed to.

Avco Corporation now had a new product line: re-entry vehicles for ballistic missiles. The firm put up a large engineering and production facility in Wilmington, Massachusetts, and that division of Avco eventually built all the re-entry vehicles for the Air Force's Minuteman II missiles and the heat shields for the Apollo space vehicles that carried astronauts to the moon and back. Kantrowitz and his research team went on to other things.

"I don't want to be merely competitive," Kantrowitz often said. "I want to be far ahead of any possible competition."

At first, Avco Everett devoted much of its efforts to developing the basic scientific knowledge needed to design better re-entry vehicles. Soon the laboratory became involved in the two-edged business of missile defense and countermeasures against possible enemy defenses. Defensive studies were called, in those days, BMD (ballistic missile defense) and, later, ABM (antiballistic missile). Offensive studies were called simply *penetration*. Both efforts hinged on the means to detect an incoming re-entry vehicle and, once it was detected, to intercept and destroy it.

Picture the situation. A missile stands in its underground silo. It is some eighty feet tall and weighs more than 100,000 pounds when fully fueled. Inside its nose cap are from three to ten re-entry vehicles, each bearing a warhead of several hundred kilotons: a hydrogen bomb with the explosive power of several hundred thousand tons of TNT. The atomic bomb that destroyed Hiroshima was estimated to have released between 15 and 20 kilotons of explosive force. The United States has a few missiles that carry warheads of 9 *megatons* — nine million tons of TNT equivalent. The Russians reportedly have 10- and even 20-megaton warheads on some of their missiles.

The missile is launched in a roar of deafening power. Straight up it climbs for a few moments, its rocket engines bellowing flame. Then it begins to arc over as it rises, gaining speed rapidly as its heavy load of propellants is gobbled up by the blazing rocket engines. In scant minutes it is higher than any airplane can fly, above the clouds and winds of earth, into the twilight realm where the sky is black except for the thin band of brilliant blue that hugs the curve of the earth's immense bulk. The first stage of the rocket booster, its propellant tanks empty and its rocket engines exhausted, falls away as the smaller second-stage engines blaze into life. But there is no sound in the vacuum of space; the missile flies on its deadly course in total silence.

Up at the top of the missile, inside the nose cap where its re-entry vehicles sit on a carrying structure called the "bus," complex electronic guidance and control mechanisms check the flight's progress. The nose cap is ejected, snapping open like the halves of a clamshell, exposing the re-entry vehicles to the vacuum and hard radiation of space. The third and last of the rocket stages burns out and falls away. The bus rides alone now, coasting across the roof of the world, heading for its rendezvous with destruction. At precisely the correct millisecond, the control system detaches the re-entry vehicles. They move out on their own now, each of them programmed to swoop in on a different target: an airbase, a missile silo, a shipyard, a hydroelectric dam, a communications center, a factory complex, a city. Each of them is marked for obliteration. The re-entry vehicles dip down into the atmosphere once more, drawn by the inexorable forces of gravity and momentum. They streak toward their targets like falling stars, leaving trails of glowing hot gases in their wakes.

How can you shoot down a six-foot-long object moving at better than ten times the speed of sound before it explodes its megaton warhead on you? It is possible to hit a re-entry vehicle with an antiballistic missile missile. ABM missiles work — but not well enough. An ABM missile can find a re-entering warhead, lock onto it, and hit it. But the attacker is not going to lob warheads at targets one at a time.

A real nuclear missile engagement will be more like the

frantic confusion of a thousand video games all going at once. The defender sits on the ground near the target with his radars and ABM missiles. The attack begins: early-warning sensors tell the defender that the attacker has launched his missiles, thousands of miles away. For the better part of half an hour, the defender can do nothing at all — except pray.

Finally the attacker's missiles begin to show up on the defender's radar screens. But instead of neat little blips that indicate where the warheads are, the screens show a wild hash created by the warheads themselves, plus thousands of decoys, cheap fakes that look to radar exactly like the bomb-carrying warheads. The attacker may also have scattered radar-confusing chaff — like the strips of aluminum foil used in World War II — to muddy the waters even further. And he may also have exploded the upper stages of his spent rocket boosters to add their fragments to the radar picture. There are so many objects cluttering the radar screen that the defender does not have enough ABM missiles to deal with them all. So the defender waits. He knows that the decoys — the chaff and tank fragments and other junk coming at him — are small and light compared with the bomb-carrying warheads. When they all re-enter the atmosphere, the light stuff will slow down or burn up, but the warheads will plow through the air without hindrance. So the defender waits for the earth's atmosphere to sort out the dangerous warheads from the confusing clutter.

That gives the defender less than a few minutes in which he can identify the warheads and launch ABM missiles to destroy them. This process is called *discrimination,* and it was the subject of intense study at Avco Everett and many other places for decades. In a nuclear engagement, every extra second the defense has is precious — literally the difference between life and death. If you can identify the warheads quickly enough, discriminate between them and the decoys and other junk, you have a chance to survive. If not, you become part of a mushroom cloud of radioactive debris.

But the attacker has other tricks at his disposal. He can explode a warhead high up in the ionosphere, up where the ozone layer exists. The hot nuclear fireball will create a vast cloud of excited electrons that form a wall impenetrable by radar waves. The defender is blinded, precisely at the critical moment when he must discriminate between the incoming warheads and the decoys. And even if the defender decides to launch his ABM missiles blindly, those missiles themselves are armed with small nuclear warheads. *Their* fireballs will add to the chaos.

In essence, the defense is in the situation of a football team that is backed up against its own goal line, desperately trying to stave off the opposition's attack. But the offense has hundreds of downs in which to put the ball across that goal line, and the game is truly in a "sudden death" situation. That is the way ballistic missile defense looked in the early 1960s. Twenty years later, it still looks that way, give or take an inch. There is no feasible way to defend targets against a determined nuclear ballistic missile attack, no matter whether those missiles are launched from underground silos or from submerged submarines, as long as the defense is restricted to sitting on the ground alongside the intended targets. The defense's position is so untenable, in fact, that the United States gave up on its ABM system — variously called Sentinel and Safeguard — in the 1970s, largely because it became apparent that such a system could not actually defend either our cities or our missile silos.

The defensive situation changes enormously, and for the better, however, if ABM weapons can be placed in space. From an orbiting satellite, the defense can shoot at the attacker's missiles almost from the moment they are launched. The ball moves from the defender's goal line to midfield, and even into the attacker's home territory, if the defense has orbital weapons. And those weapons can be non-nuclear, an immense advantage both to the defender and to the health and welfare of the earth.

We can replay the missile attack scenario now, adding ABM satellites to the defender's side. Picture a hundred such satellites crisscrossing the earth in low orbits, two hundred miles up or less. The satellites are armed with either lasers or particle-beam devices. Lasers emit pure energy, intense beams of visible or infrared light. Particle-beam devices are like the "atom-smasher" accelerators that physicists use; they produce streams of very energetic atomic particles, such as protons and electrons. Both types of devices are called *energy-beam weapons,* or simply beam weapons when used against military targets. No matter whether it fires a beam of light or a stream of particles, a beam weapon strikes with the speed of light, 186,000 miles per second. Nothing in the universe moves faster. Beam weapons are literally the fastest gun in town. By comparison, the Mach 10-plus speed of an ICBM seems turtlelike.

With a hundred or more beam-weapon satellites in low orbit, the defense can cover any spot on earth twenty-four hours per day. No matter when or where an attack is launched, there will be enough satellites with enough fire-power to destroy the attacking missiles.

As soon as the missiles are launched, sensors aboard early-warning satellites alert the defense. The ABM satellites turn their beam weapons to meet the rising missiles and lock their pointing and tracking systems onto the targets. The missiles are easy to spot as they climb up from the surface, bellowing bright tongues of flame from their rocket engines. They are also rather fragile at this stage of their flight. As the missiles clear the dense lower layers of the atmosphere, energy beams lance out from the satellites, crossing the hundreds or thousands of miles of space to their targets almost instantaneously. The intense beams slice through the thin skins of the missiles' propellant tanks. The missiles begin to explode, bright blossoms of chemical fire dotting the dark sky.

Some missiles survive the first salvos from the satellites, drop their booster stages, and light up their second-stage rocket engines. But as they continue on their preset courses, they come within range of other satellites that were originally too far away to fire at them. Now these fresh satellites form a second line of defense, hammering away at the missiles while their upper-stage engines are burning, or even afterward, when they have started their long coasting flight. Energy beams slash at the missiles' upper stages, at the warhead-carrying buses, even at the warheads themselves. The warheads are built to withstand the heat and shock of re-entry; they would be the hardest targets to destroy. Far easier to knock out the rocket boosters so that the warheads never reach their targets. Still, a determined attack by energy beams might damage even the toughest re-entry vehicles sufficiently so that they will not be able to detonate their nuclear warheads even if they get through to their targets.

Up until the mid-1960s the idea of energy-beam weapons was strictly fantasy. But Kantrowitz and his team of bright young men changed all that. Because the laboratory dealt mainly with the physics of very hot gases, and because we were not bashful about telling the world how good we were, Avco Everett staffers became known through the aerospace industry as "hot-air specialists." We did a lot of very interesting things with "hot air." In addition to our continuing work on re-entry physics, missile defense, and penetration, we studied the fundamental behavior of ionized gases and their interactions with magnetic fields. This area of research was called *magnetohydrodynamics,* or MHD, for short. The lab developed the first MHD electric power generators and investigated the possibilities of using MHD forces to propel spacecraft. Kantrowitz and his brother, a nationally prominent thoracic surgeon, worked together on several medical devices, including some of the earliest artificial hearts.

Then, in 1960, the world's first laser flashed its ruby-red light at the Hughes Research Laboratories in Malibu, California. Within a few years laboratories all around the industrialized world were building lasers that used various gases as their working mediums: neon, argon, carbon dioxide, helium, nitrogen lasers, emitted light beams of every color in the rainbow, as well as infrared and ultraviolet beams, invisible to the human eye.

Kantrowitz, always seeking to produce results "that have an effect on the world," challenged his staff to invent a more powerful laser. He wanted a device that could become a useful tool, rather than a laboratory curiosity. By 1965 Avco Everett had succeeded. The physicists invented the *gasdynamic* laser. Until then, gas lasers were all built around a container that held the gaseous working medium: a glass tube, usually, that was surrounded by the mirrors and electrical components necessary to make the laser emit energy. I remember visiting the Redstone Arsenal, in Alabama, in the mid-1960s, where Army researchers had built a carbon dioxide laser consisting of a slim glass tube 178 feet long. It emitted a 3000-watt invisible beam of infrared energy, the most powerful laser ever built — until then.

The Avco Everett team soon realized that what was preventing lasers from achieving higher power outputs was what they called "garbage removal." In the ordinary gas laser, the working gas heated up as it emitted energy, and quickly exhausted its reservoir of energy. Power emission fell off until the gas was resupplied with energy and the heat and waste products it had accumulated trickled away.

But what if the working gas were not confined to a tube? What if the laser were built somewhat like a wind tunnel, where the used-up working gas flows out of the laser area and fresh new gas is brought in, at supersonic speed? Instead of a septic tank for garbage removal, the laser now has a high-speed sewer line. The first gasdynamic laser, with a working section no bigger than an ordinary cigar box, produced more power output during its first ten seconds of operation than all the lasers built since the first one, in 1960, combined.

The gasdynamic laser was more akin to a flame-thrower than a flashlight. It was destined to make an impact on the world.

In 1966 Avco Everett briefed the Pentagon about the breakthrough we had achieved. Why the Pentagon? Certainly high-power lasers have many potential applications outside the military: cutting and welding metals, driving tunnels through hard rock, interplanetary communications, drilling for oil or natural gas. But their military applications are obvious. Weapons are nothing more than tools for delivering energy onto a target. The more energy that can be unleashed on the target, and the greater the distance over which the energy can be wielded, the more effective the weapon. That is why a club "outvotes" a fist, why a pistol is more lethal than a knife.

In 1966 it was apparent to us that high-power lasers had vital military implications. They had the potential to deliver enormous amounts of pure energy over extremely long distances. We were patriotic (or chauvinistic) enough to feel that such weapons potential ought to be developed for the defense of the United States. We knew, from our long years of investigating missile re-entry, that the nuclear-armed ballistic missile was still regarded as the "ultimate weapon." And justly so. But high-power lasers offered a glimmer of hope for the defense; perhaps they could be developed to defend us against nuclear annihilation.

We knew, too, that as soon as Washington learned about the gasdynamic laser, the concept would be placed under military security. Yet there is no way to keep nature secret: if we could discover the breakthrough to truly high-power lasers, someone else would, sooner or later.

In fact, we watched with a mixture of curiosity and apprehension as the Russian scientific community followed the same trail we had. There was no espionage involved, no spies filching secrets in the dark of night. Molecules work the same way for Communists as they do for capitalists. At that time the Soviets did not classify theoretical scientific research, so we could follow the Russian technical journals as different groups of scientists pursued the three main avenues of investigation that led toward the realization that very high-power lasers could be built. Precisely at that point, someone in the Soviet Union gave out a resounding "Aha!" and all research along those lines disappeared from the open, unclassified literature. The Russians were about a year or two behind us in inventing the gasdynamic laser. They worked very hard to catch up, and even to surpass us, just as they did with nuclear weaponry.

Moreover, the Pentagon was then the major source of funding for research in the physical sciences. Even today, after the horror and revulsion of Vietnam and the massive protests of the Nuclear Freeze Movement and other peace movements, most of the nation's research and development in the physical sciences is still funded by the Department of Defense. The Pentagon literally has a thousand dollars to spend on R and D for every dollar that other funding sources have.

That first briefing in Washington was snowed out by a

January blizzard, even though I dutifully carried the top secret slides for our presentation from Boston to the capital — by train. The ride took nearly fifteen hours, and it was like a scene from the Trans-Siberian Railroad. It was past midnight by the time I arrived in Washington, cold and hungry and lugging an enormous locked valise full of classified slides. Nothing was moving through the two-foot snowdrifts that covered the city. I had to mush my way on foot from the train station to the Statler Hilton Hotel, only to learn that the meeting had, of course, been canceled.

Several weeks later, in February 1966, Kantrowitz and several of his key scientists spent more than an hour in a Pentagon conference room presenting their story about the gasdynamic laser. But not before we had scoured the halls for a technical sergeant who could fix the slide projector we were to use; it was faulty, and none of the multimillion dollars' worth of scientific talent sitting around that conference table could fix it. The sergeant was not cleared for our top secret material, so we had to usher him out of the room after he had fixed the balky projector.

When the last slide had been shown, the cranky projector turned off, and the ceiling lights switched on again, there was a long, dramatic moment of awed silence in that conference room. The best experts that the Defense Department could call on, top men in the fields of electronics, optics, and physics, had been stunned by our presentation.

Within months we were not only developing the technology of high-power lasers; we were also participating in studies of how they could be used. One of the earliest such studies was code-named Eighth Card. That and other Card-titled studies convinced me that there were some pretty serious poker players at high levels in the Pentagon. These studies pointed to the possibility of using lasers as a defense against ballistic missiles. But it became clear that the lasers would have to be placed in space, where there was no air to absorb or distort their beams of energy, and where they could shoot at the attacker's missiles almost as soon as they were launched.

The power outputs of such ABM lasers would have to be enormous, millions or even billions of watts. In the jargon of the technologists, megawatts or gigawatts. The highest power levels of today's lasers are classified secret by the Defense Department. But in 1983 the Air Force released news that a 400-kilowatt (400,000 watts) laser flown aboard the Air Force's Airborne Laser Laboratory — a specially outfitted C-135 cargo jet — had successfully shot down five Sidewinder missiles fired at it by a jet fighter plane. The test took place high above the Navy Weapons Center at China Lake, California. Sidewinder missiles are the kind that fighter planes use to destroy other planes: air-to-air missiles. Although the laser did not destroy the Sidewinders, it damaged their heat-seeking sensors so severely that the missiles could not find their target and crashed into the desert.

Meanwhile, TRW Corporation has built a laser of 2.2-megawatt output for the U.S. Navy. Although it is not intended to fly, this laser is getting into the power range of interest for orbital ABM weaponry. It is called MIRACL, a somewhat whimsical acronym for midinfrared advanced chemical laser. Installed at the White Sands Missile Range in New Mexico, it is used by all three armed services to study the mechanisms by which laser energy damages target materials, such as the metals of which aircraft and missiles are constructed.

I left Avco Everett Research Laboratory in 1971, to re-enter the publishing industry as the editor of *Analog Science Fiction–Science Fact* magazine. Later I became the editor of *Omni* magazine. All during the years I was with Avco, and during my years as an editor, I continued to write books, and I have thought about and studied the many facets of ballistic missile defense for more than twenty years. I have worked with the scientists, engineers, military officers, politicians, bureaucrats, and futurists. In such novels as *Kinsman* and *Millennium* I have portrayed how these people think and feel, and how they are changing the world around us. In my nonfiction books, such as *The High Road* and *The Fourth State of Matter*, I have shown how these new technologies work and where they may be leading us.

When President Reagan's speech made it apparent that Washington is pursuing orbital ABM defenses, based primarily on high-power lasers, I realized that I had to share what I knew about the subject and the conclusions that I have reached after more than two decades of thinking and writing about it.

My conclusions are quite straightforward.

1. It will be possible to orbit satellites armed with energy-beam or other types of weapons that can destroy ballistic missiles long before they reach their targets.
2. Placing such satellites in orbit could seriously destabilize the existing "balance of mutual terror" that has characterized East-West relations for the past twenty-odd years. The ABM satellites could well trigger the nuclear war they are designed to prevent.
3. Therefore, we have, almost within our grasp, the *technical* means to prevent a nuclear holocaust. But we lack the *political* means to use it. It will take at least ten years before this technology is fully developed and ready for use. We have ten years, then, in which to decide how — or even if — we want to use it.

In the chapters that follow I will try to show how we can create the political means to protect ourselves against nuclear Armageddon. I will use my skills as a novelist and my long friendships with many of the key personalities involved in high technology and politics to draw scenarios, minidramas peopled with fictitious (but quite realistic) characters, in addition to chapters that are solidly historical and others that are frankly speculative. My intention is to reveal the *human* thinking and motivations that create and guide the politics and technology.

You will meet many people in the coming chapters of this book. You will hear many different voices. Some of them will be annoying, troubling, demanding. Many new ideas, and new interpretations of well-known events, will be played out for you. In the end, I want to leave you with enough information, both technical and political, so that you can make up your own mind about how we should utilize the new capabilities that space technology is offering us.

Make no mistake: we are talking about life and death, not merely for ourselves as individuals, but for our society, our whole world, the future of the entire human race. The decisions we make over the next decade will determine whether or not humankind survives to see the twenty-first century.

We shall begin with a brief look into the recent past, because, as Patrick Henry put it, "I have but one lamp by which my feet are guided; and that is the lamp of experience. I know of no way of judging the future but by the past."

PART II

The Road to Now

History is shaped by the time in which it is written.
— Daniel Yergin

4

Countdown to Armageddon

SOMEWHERE IN THE WORLD there are slightly more than 4 tons of explosive energy waiting to kill you. Between 4 and 4.5 tons of nuclear death waits in some nation's arsenal, with your individual name on it.

There are an estimated fifty thousand nuclear warheads in the world today. Some sit atop ballistic missiles waiting silently in their underground silos or cruising deep beneath the ocean's surface in nuclear submarines. Others are in bombs carried by airplanes and cruise missiles. Still more wait in heavily guarded storage areas in Europe and Asia, waiting to be "mated" to medium-range and short-range missiles for battlefield use.

These warheads have an aggregate explosive power of roughly twenty billion tons of TNT: 20 gigatons. That equals some 4 to 4.5 tons of destruction for each man, woman, and child on earth.

How did the world get into this terrifying situation, and why? To understand the problems and possibilities of warfare suppression, it is necessary to understand the four-decade-long arms race that has brought the world to the brink of nuclear oblivion.

3 October 1942

The V-2 rocket missile designed by Wernher von Braun and his team of engineers and technicians at Peenemünde, on the Baltic coast of Germany, makes its first successful test flight. Its two earlier tests were failures, although in the second flight the missile did exceed the speed of sound before going off course and crashing.

8 September 1944

The first V-2 missiles are fired at southern England. Two days earlier, a pair of V-2s was fired at Paris, but never reached the target. Over the next seven months, nearly 4500 V-2s are fired at England and targets in the Low Countries. Some 1500 hit London and southern England, killing more than 2500 people; 1610 are launched against Antwerp.

The V-2 is slightly more than forty-six feet long, sixty-five inches in diameter, and weighs more than twenty-seven thousand pounds. It carries a 2200-pound warhead of chemical explosive over a maximum range of 210 miles in a total flight time of five minutes. It dives at its target at a speed of 3500 miles per hour, too fast to be seen, heard, or intercepted. The only defense against the V-2 is to destroy or capture its launching platforms.

1945
Franklin D. Roosevelt dies, 12 April; Harry S Truman becomes U.S. President

1945
Germany surrenders,
7 May

16 July 1945

The first atomic bomb is exploded at Alamogordo, New Mexico, in a test code-named Trinity. Although theoretical calculations predicted an explosion yielding an energy release of some 5000 tons of TNT, the Trinity bomb releases between 15,000 and 20,000 tons of explosive energy: in the parlance of the scientists, 15 to 20 *kilotons*.

J. Robert Oppenheimer, head of the Los Alamos laboratory, which built the bomb, watches the fireball searing the New Mexico morning and thinks of lines from the Bhagavad Gita: ". . . the light of a thousand suns suddenly arose in the sky. . . . I have become Death, destroyer of worlds."

24 July 1945

President Harry S Truman, at the Potsdam Conference in suburban Berlin, hints to Marshal Joseph Stalin that the United States has developed "a new weapon of unusual destructive force." Stalin, already informed by Soviet intelligence agents of the United States' Manhattan Project, orders a speedup of the Russian atomic bomb work immediately. Russian scientists had been working in nuclear physics since the 1920s. Igor Kurchatov was the first to predict, in 1939, that an atomic bomb could be made by using uranium fission.

6 August 1945

Hiroshima is destroyed by an atomic bomb, dropped from the B-29 *Enola Gay*. The uranium-fission bomb releases an explosive force estimated to be equal to nearly 20,000 tons of TNT. Some sixty-eight thousand people are killed instantly; another ten thousand are missing and never found; thirty-seven thousand are wounded. More than fifty thousand people are affected by radiation sickness in following years.

9 August 1945

Nagasaki, a port city of 250,000, is severely damaged by a plutonium-fission bomb, dropped from a B-29 nicknamed *Bock's Car*. Although the explosive force of the "Fat Man" bomb (so called because its interior design gave it a girth of more than fifteen feet) is somewhat higher than the Hiroshima weapon, casualties are lower, because Nagasaki's hilly terrain restricts the force of the blast and radiation. About thirty-eight thousand people are killed immediately; a like number are wounded and injured by radiation.

June 1946

Bernard Baruch presents the so-called Baruch Plan to the United Nations. Based on studies done originally by Dean Acheson and David Lilienthal, the plan outlines an attempt

1945
Japan surrenders, 14 August

1946
United Nations General Assembly holds its first session, 7 January

to control the development of nuclear power and prevent the proliferation of nuclear weapons. It calls for an International Atomic Energy Authority that would acquire complete control over all raw materials and manufacturing plants for all types of nuclear developments, worldwide. A few days later, Andrei Gromyko presents the alternative plan of the Soviet Union, which calls for banning the production and use of nuclear weapons, and destruction of all such weapons now in existence, without international inspection or controls.

Thus, the American plan would have the Russians open their nuclear laboratories to international inspection and control while the Soviet plan would demolish the American arsenal of nuclear weapons — the only such arsenal in existence at that time. By 17 September, Baruch reports to President Truman that the negotiations are deadlocked. Neither plan is approved, and the United Nations loses its first, perhaps its only, chance to gain control over nuclear weaponry.

1946
Communist regimes established in Poland, Rumania, Hungary, Bulgaria, Yugoslavia, Albania, East Germany

15 March 1947

The Soviet government authorizes formation of a state commission to examine the feasibility of developing long-range ballistic missiles. The commission is to report directly to Joseph Stalin, who remarks that an intercontinental ballistic missile "could be an effective straitjacket for that noisy shopkeeper Truman. We must go ahead with it, comrades! The problem of the creation of transcontinental rockets is of extreme importance to us."

Since the early 1930s Russian rocketry enthusiasts had been experimenting with liquid-fueled and solid-propelled rockets. During World War II, rocket work was accelerated, with the emphasis on short-range missiles for battlefield use. Captured German rocket scientists are kept separate from the Russian rocket teams and are used mainly as independent sources to check on the progress made by the Russians themselves.

June 1947

A U.S. Department of Defense study, "Operational Requirements for Guided Missiles," draws the conclusion that long-range ballistic missiles are not feasible, and commands the U.S. Air Force to concentrate its efforts on developing pilotless aircraft rather than rocket-powered ICBMs. The fledgling American rocket industry is virtually destroyed by this decision, which is based in part on the erroneous assumption that nuclear weapons will

1947
U.S. Army reduced from 1945 level of more than 8 million to 690,000

Soviet Red Army reduced from some 11 million to 2.8 million

1948
U.S. Congress passes $17 billion Marshall Plan for economic aid to Europe

Communist coup d'état in Czechoslovakia, 25 February

Berlin blockade begins, 24 June. U.S. counters with airlift

Nation of Israel created; first Arab-Israeli war

always be too large to be carried by ballistic missiles.

Rocketry in the United States is reduced to a low-paced scientific research effort; in the Soviet Union ballistic missiles are being developed as rapidly as possible. Thus, the Russians gain a headstart of almost ten years in the development of large rocket boosters.

*1949
Berlin blockade ends, 4 May. U.S. airlift continues until September*

Communists win civil war in China

NATO pact signed in Washington

*1950
North Korean forces invade South Korea, 25 June*

29 August 1949

The Soviet Union explodes its first atomic bomb, ending the American monopoly on nuclear weapons. American strategists, who had been undecided about developing a hydrogen-fusion, or thermonuclear, "super" bomb, recommend to President Truman that the super project should be initiated. Robert Oppenheimer, who had opposed development of the hydrogen bomb, mainly on moral grounds, four years later has his security clearance taken away from him and leaves government service.

3 October 1952

Britain detonates its first atomic bomb in a test on the Monte Bello Islands, off Western Australia, thus becoming the third nation to join the "nuclear club."

1952
Dwight D. Eisenhower elected U.S. President

1 November 1952

The United States explodes its first hydrogen bomb in the Ivy Mike test at Eniwetok Atoll in the central Pacific Ocean. A 65-ton experimental device, Ivy Mike releases more than ten million tons of TNT in explosive power (10 megatons). The explosion blows a one-mile-wide crater that obliterates the islet of Elugelab. The crater is about fifteen hundred feet deep, roughly as deep as the height of the Empire State Building. Some intelligence information indicates that the Soviet Union performed a similar test several months before the Ivy Mike shot, but this is never confirmed, nor is the information widely disseminated, even within the scientific and military nuclear weapons communities.

1953
Joseph Stalin dies, 5 March. Nikita Khrushchev appointed First Secretary of Communist Party Central Committee; G. P. Malenkov named USSR Premier

Korean armistice signed, 27 July

12 August 1953

The Soviet Union explodes its first "deliverable" hydrogen bomb, a compact device capable of being carried by an airplane or missile. Slightly more than six months later, 1 March 1954, the United States explodes a deliverable H-bomb in the Bravo test at Bikini Atoll. Japanese fishermen aboard *Lucky Dragon*, 75 miles from the test site, and Micronesian natives on the island of Rongerik, 125 miles away, receive lethal doses of radioactive fallout.

1955
Nikolai A. Bulganin succeeds Malenkov as USSR Premier

21 July 1955

President Dwight D. Eisenhower stuns the Soviet delegation to the Geneva Summit Conference — and many members of his own delegation — by proposing an Open Skies policy to prevent surprise nuclear attack.

The conference, convened to discuss disarmament, quickly bogs down into the usual Cold War polemics. Frustrated, Eisenhower speaks directly and without notes to Soviet Premier Nikolai Bulganin and Communist Party Secretary Nikita Khrushchev, offering to exchange the complete military blueprint of all American armed forces for those of the Soviet Union, and suggesting regular and frequent inspection of each nation's military installations by aircraft overflights. He bases his proposal on studies done by a panel of experts chaired by Nelson Rockefeller.

"I do not know how I could convince you of our sincerity in this matter, and that we mean you no harm," Eisenhower tells the Russians. "I only wish that God would give me some means of convincing you of our sincerity and loyalty in making this proposal." At that moment, a flash of lightning from the thunderstorm raging outside the conference hall knocks out the electrical power and

1956
November — Hungarian revolt crushed by Soviet troops

Suez war involves Israel, Egypt, France, Britain

plunges the meeting into darkness for several minutes.

Bulganin seems genuinely moved by Eisenhower's personal appeal. But that evening, Khrushchev asks Eisenhower, "Who are you trying to fool? In our eyes this is a very transparent espionage device. . . . You can hardly expect us to take this seriously."

15 May 1957

Great Britain joins the United States and Soviet Russia by exploding its first hydrogen bomb, estimated to have the power of 1 megaton of TNT, in a test near Christmas Island, some twelve hundred miles south of Hawaii. Sir Anthony Eden, the British Prime Minister, announces that the weapon was dropped from a Vulcan jet bomber flying at an altitude of forty-five thousand feet and was detonated at fifteen thousand feet above the surface of the Pacific so that radioactive fallout would be minimized. The Japanese government expresses "strong regret" at the British decision to test hydrogen bombs. The British Labour Party had asked the government to postpone nuclear testing as a gesture toward disarmament, but Eden's Conservatives countered that Britain could take a more effective stance in disarmament negotiations once it had its own hydrogen bomb.

August 1957

The Soviet Union test-flies its first ICBM, an SS-6, over a four-thousand-mile range. In April 1956 the Russians had flown their first intermediate-range ballistic missile (IRBM), a shorter-range rocket capable of reaching the major cities of Western Europe from launching sites within the USSR. Earlier test flights, observed by Western intelligence radars, convinced the U.S. Department of Defense that ballistic missiles were indeed feasible. A crash program to catch up to the Russians is initiated in the United States, involving, at first, all three military services, although ultimately the Air Force wins responsibility for ground-based ICBMs, the Navy is assigned to develop submarine-launched ballistic missiles, and the Army is given the task of ballistic missile defense.

Meanwhile, Nikita Khrushchev creates a new branch of the Soviet armed forces, the Strategic Rocket Forces, and tells his generals that ballistic missiles will allow him to reduce the size of the Red Army. By 1958 the army is shrunk from 5.7 million men to slightly more than 3.6; but it begins to grow again in 1961, with the Berlin Wall crisis and the combative rhetoric of the new Kennedy administration.

4 October 1957

The Soviet Union launches the world's first artificial satellite, Sputnik I, using a modified SS-6 ICBM. Over the next several years the USSR achieves an unbroken string of space "firsts," including placing the first animals in orbit, sending a spacecraft into orbit around the sun, and photographing the far side of the moon, which is never seen from earth.

6 December 1957

The U.S. space program flounders as Vanguard, the official American satellite program, suffers a disastrous launch failure. The first attempt to launch a Vanguard satellite ends just above the launch stand, as the rocket's first-stage engine explodes.

31 January 1958

Von Braun's team from the U.S. Army's Redstone Arsenal, in Huntsville, Alabama, launches the first American satellite, Explorer I. The launching vehicle is a modified Jupiter IRBM, developed for the Army by von Braun and his German engineers and technicians.

Easter Sunday 1958

The philosopher Bertrand Russell, at the age of eighty-six, begins to devote most of his time to efforts in Britain and elsewhere to nuclear disarmament. In writings, speeches, letters to

Khrushchev and Eisenhower, and eventually through the medium of huge Ban the Bomb demonstrations, Russell and other leaders of the Committee for Nuclear Disarmament give voice to the general public's fear of growing nuclear arsenals and East-West hostility. Although in 1945 Russell foresaw a world in which the United States used its nuclear monopoly to enforce global peace and democracy, by 1959 he helps to stage a debate in the House of Lords on a motion to have Britain dismantle its own nuclear weapons and urge the non-nuclear nations to renounce their manufacture and use. As part of the so-called Peace Movement, the efforts of Russell and others lead toward the SALT and Nonproliferation treaties of the 1970s — and to the Nuclear Freeze Movement of the 1980s.

31 October 1958

After several years of diplomatic maneuvering and negotiation, and under increasing public pressure to stop the radioactive fallout from nuclear bomb tests, the United States and the Soviet Union agree informally to a moratorium on testing nuclear weapons. The moratorium lasts almost three years, but is broken suddenly in 1961, when the Russians resume testing without warning.

1959
Fidel Castro takes over government of Cuba

23 December 1958
The first American ICBM, an Atlas C, completes a full-range flight test of four thousand miles. Later developments stretch the range of the Atlas to nine thousand miles. The U.S. Air Force also develops the Titan ICBM. Both missiles are liquid-fueled; the Titan, however, uses propellants that can be stored in the missile's tanks for long periods, so the missile can be ready for firing in a relatively short time.

13 February 1960
France successfully tests its first atomic bomb, becoming the fourth member of the nuclear club. The plutonium bomb is exploded near Reggan in the Sahara of western Algeria; it is estimated to have an explosive yield of about 20 kilotons. French President Charles de Gaulle says, "Hurrah for France! Since this morning she is stronger and prouder."

Russian and American representatives in Geneva, who are trying to draw up a mutually acceptable ban on testing nuclear weapons, react with much less enthusiasm. In Washington, the official White House statement is restrained, neither surprised nor pleased. Moscow voices regret at the French decision to "go nuclear," repeats its call for a complete ban on all testing, but says that if the French arm themselves with nuclear weapons, the Soviet Union "will draw relevant conclusions."

20 July 1960

The nuclear submarine *George Washington* successfully launches a Polaris SLBM while submerged, almost three years before the Navy's original deadline for such a test. By November 1960 the *George Washington* ships out into the Atlantic on its first operational cruise, carrying sixteen Polaris missiles — the world's first ballistic-missile-carrying submarine.

The U.S. Navy's Polaris program forms the third leg of the American strategic triad: land-based ballistic missiles, manned bomber aircraft, and submarine-based ballistic missiles. The triad is seen by American strategists as maximizing the nation's flexibility in delivering a nuclear attack while at the same time greatly complicating the Soviets' ability to respond to the three-legged threat.

11 August 1960

An Air Force transport plane makes a midair "catch" over the Pacific Ocean of a photographic film cassette ejected by Discoverer 13 from orbit. Although the film cassette was empty, the successful test opens a new era in the use of satellites for surveillance.

Coming a scant few months after Russia's downing of a U-2 intelligence plane, the satellite moves photographic reconnaissance and surveillance to a new level. The Soviet Union protests against satellite overflight of its territory, but since the Russians themselves established the precedent of overflying without permission when they launched Sputnik I, their protests are ignored. Within a few years, both the United States and the Soviet Union are regularly launching observation satellites, which count missile silos and submarines, and disarmament agreements are eventually based on data obtained by satellite cameras, which the diplomats refer to as "national technical means of verification."

8 November 1960

John F. Kennedy is elected thirty-fifth President of the United States. During his election campaign he decried the "missile gap," based on official intelligence estimates that the Soviets had deployed hundreds of ICBMs, while the United States had only a few. Once in the White House, Kennedy discovers that the missile gap exists in reverse: although the Russians developed missiles before the United States, American production has far exceeded that of the Soviet Union. The United States has a wide advantage over the Soviets. According to John Prados, in his book *The Soviet Estimate*, by mid-1961 the U.S. Air Force had deployed seventy-eight strategic missiles, compared to "a few individual [Russian] missiles, some quite powerful, but hardly a coherent force capable of a simultaneous disarming strike against U.S. retaliatory forces."

In addition, the United States strategic striking forces included eighty submarine-based Polaris missiles and more than fifteen hundred long-range jet bombers, as compared with fewer than two hundred Soviet long-range bombers and no operational SLBMs.

1961
Berlin Wall constructed

Bay of Pigs invasion of Cuba fails

1 February 1961
The first test flight of the Minuteman ICBM is a complete success. This three-stage rocket uses solid propellants, which allows the missile to be stored indefinitely in underground silos yet still be ready for launching at almost an instant's notice. Smaller than the Titan ICBM, and carrying a lighter warhead, Minuteman nevertheless is much less expensive than the Titan and can be manufactured and deployed in much greater numbers. By early 1963 it is operational and quickly becomes the backbone of the U.S. ICBM force.

12 April 1961

Lieutenant Yuri Gagarin, of Soviet Russia, is the first human being to fly in space. His Vostok capsule nearly circles the earth in a flight that ranges from 113 to 204 miles' altitude. After 108 minutes in space, his craft re-enters and is recovered in the Ternov District, between the Volga River and the Ural Mountains.

1 September 1961

Soviet Russia resumes nuclear testing in the atmosphere, ending the informal test moratorium that had begun three years earlier. The USSR conducts thirty tests, including a 60-megaton behemoth, before the United States can resume testing. The nuclear tests are seen in the West as part of a new Soviet effort to overtake the United States in strategic striking power. The arms race is accelerating, most experts agree.

20 February 1962

Lieutenant Colonel John H. Glenn is the first American to orbit the earth, in a Mercury capsule launched atop a modified Atlas ICBM. His three-orbit mission lasts four hours and fifty-five minutes. He is recovered at sea, 210 miles northwest of San Juan, Puerto Rico.

October 1962

The United States and Soviet Russia teeter on the brink of nuclear war over the Cuban missile crisis. The crisis begins when U.S. intelligence flights discover Soviet missile bases being built in Cuba. A naval blockade of Cuba prevents the USSR from arming the missiles with nuclear warheads. The crisis is resolved peacefully, but not before the strategic nuclear forces of both sides go on wartime alert status.

The effects of the Cuban missile crisis are far-reaching. Washington and Moscow, aware of how close to war the two superpowers were, both engage in serious efforts to move away from nuclear confrontation. The era eventually known as détente begins. But Khrushchev, personally humiliated by the forced Russian backdown in Cuba, and lacking the support of the Red Army (in part because of his insistence that missiles have made the army less important), is soon removed from power.

23 September 1963

After barely a month of debate, the U.S. Senate ratifies the Limited Nuclear Test Ban Treaty by a vote of 79–9. The treaty is a major step in détente.

When the chief of the American delegation, W. Averell Harriman, arrived in Moscow in the summer of 1963 to begin negotiations on the treaty, reporters asked him how long the negotiations would take. "About two weeks," he replied, "if Premier Khrushchev wants a treaty as badly as President Kennedy does." Thirteen days later the treaty was drafted and approved by the Russian and American negotiators.

The treaty prohibits the testing of nuclear weapons in the atmosphere, under the ocean's surface, or in space. For years negotiators had tried, unsuccessfully, to draft a comprehensive treaty banning *all* nuclear weapons testing. The Limited Test Ban Treaty reflects both growing public concern over the rising amounts of radioactive fallout from nuclear tests in the atmosphere and the worldwide fears of nuclear war that had been stirred by the Cuban missile crisis.

Much of the debate in the Senate revolves around the possibility that the Soviet Union might be able to surpass American nuclear know-how by conducting weapons' tests underground that could not be detected by the United States. Herbert York, a top Department of Defense scientist during the Eisenhower and Kennedy administrations, and Jerome Wiesner, Kennedy's Science Adviser, write in *Scientific American* that Soviet under-

1963
President Kennedy assassinated, 22 November; Lyndon B. Johnson becomes U.S. President

ground tests can be detected, and that no foreseeable American weapons' developments are jeopardized by the treaty. They add:

> Paradoxically, one of the potential destabilizing elements in the present nuclear standoff is the possibility that one of the rival powers might develop a successful antimissile defense. Such a system, truly airtight and in the exclusive possession of one of the powers, would effectively nullify the deterrent force of the other, exposing the latter to a first attack against which it could not retaliate.

The Limited Test Ban Treaty eases public fears of radioactive fallout, even though France and, later, China continue to test nuclear weapons in the atmosphere. A *comprehensive* test ban treaty, which would prohibit all nuclear testing, remains an unfulfilled dream.

1964
U.S. begins massive troop buildup in South Vietnam

Khrushchev replaced as Soviet Premier by Aleksey N. Kosygin and as General Secretary of the Communist Party by Leonid I. Brezhnev

1964–1968

The U.S. Air Force tests an antisatellite (ASAT) weapon system on Johnston Island, an isolated atoll nearly a thousand miles southwest of Hawaii. After a high-altitude nuclear bomb test in 1962, code-named Starfish, knocked out the electronics on a number of orbiting satellites, scientists realized that the pulse of electromagnetic energy from a nuclear explosion in space could damage satellites over distances of thousands of miles. The ASAT tests, called Squanto Terror

(Program 437), are made with a Thor booster and dummy warheads, since the Limited Test Ban Treaty forbids nuclear explosions in space. Targets are old satellites that are no longer functioning. Within a year, the warheads are being placed less than a mile from the orbiting targets.

After sixteen tests, Program 437 is terminated, partly because the 1967 Outer Space Treaty (see below) has eased the threat of Soviet orbital nuclear bombs, and partly because a nuclear explosion in space would damage friendly satellites, as well as hostile ones, through the massive electromagnetic pulse (EMP) caused by the explosion.

1965
Start of student demonstrations in U.S. against Vietnam War

16 October 1964
The People's Republic of China becomes the fifth member of the nuclear club, exploding its first atomic bomb in the western province of Xinjiang. The bomb is composed of highly enriched uranium-235; its yield is estimated at 10 to 20 kilotons.

1967
Six Day War between Israel, Syria, Egypt, Jordan

27 January 1967
Representatives of sixty-three nations, including the United States and the Soviet Union, sign the Outer Space Treaty, which prohibits claims of national sovereignty over the moon or other celestial bodies, bans the creation of military bases on the moon,

and prohibits the placing of "weapons of mass destruction" in space. "Mass destruction" weapons are understood to include nuclear, chemical, biological, and radiation weapons.

17 June 1967
The People's Republic of China explodes its first hydrogen bomb, after five previous tests of atomic weapons, including tests in which the bombs were dropped from a plane or flown to the target aboard a ballistic missile.

3 November 1967
Defense Secretary Robert McNamara reveals that the Soviet Union is testing a fractional orbit bombardment system (FOBS). Although keeping within the letter of the Outer Space Treaty, the Russian FOBS can loft a nuclear warhead into a partial orbit, then de-orbit it over a target. Such a delivery system allows only a few minutes of warning time for the defense, since it hugs the horizon and it is unclear until the de-orbiting maneuver whether the object is an actual satellite being placed in orbit or a nuclear warhead on its way to a target on the ground.

1968
Dubček government deposed as Warsaw Pact forces invade Czechoslovakia

1968
The Soviet Union begins tests of an orbital antisatellite weapon. It is a "killer" satellite boosted into orbit aboard a modified SS-9 launching

1968
Viet Cong launches Tet offensive in 100 South Vietnamese cities and hamlets

Richard M. Nixon elected U.S. President

1970
Four students killed at Kent State University during protest against Vietnam War

1971
Fighting spreads to Laos and Cambodia

"Pentagon Papers" published by the New York Times

rocket, then maneuvered into an orbit that brings it within a mile of its target satellite, close enough so that its warhead of conventional explosive destroys both satellites. Test targets in orbits as high as twelve hundred miles are successfully intercepted and destroyed.

24 August 1968

France joins the hydrogen club by exploding its first H-bomb in a test in the South Pacific Ocean.

5 March 1970

The Nuclear Nonproliferation Treaty goes into effect, ratified by forty-three nations nearly two years after it was first prepared for signature. Nuclear powers, like the United States and the Soviet Union, agree not to assist non-nuclear nations in any way to acquire nuclear weapons. Non-nuclear nations agree not to produce nuclear weapons and to allow inspection of their nuclear facilities. Among the nations that have not signed the Nonproliferation Treaty are France, China, Israel, Pakistan, India, Cuba, Argentina, Brazil, South Africa, Saudi Arabia, and a number of other Arab states.

26 May 1972

In the Great Hall of the Kremlin, President Richard Nixon and General

1972
Arab terrorists kill 11 Israeli athletes at Olympic Games in Munich

1973
Vietnam cease-fire agreement signed; U.S. troops leave South Vietnam after 45,948 combat deaths

Arab-Israeli Yom Kippur War

Arab oil embargo against U.S., Western Europe, and Japan triggers energy crisis

Secretary Leonid Brezhnev sign the SALT I agreement and the ABM Treaty, after nearly four years of negotiations.

The Strategic Arms Limitation Treaty (SALT I) freezes the number of ICBMs and SLBMs each side can possess to the levels in operation or under construction as of 1972. Thus, the United States maintains its ICBM force at 1054 and its submarine-launched missile force at 656, levels they have remained at since roughly 1966. The Soviet Union, which has been rapidly adding to its forces since 1966, promises to maintain them at 1618 silo-based missiles and 710 submarine missiles. SALT I does not cover bombers or the number of warheads each missile may carry.

Because of the technique of the multiple independently targeted re-entry vehicle (MIRV), whereby one missile can carry three or more warheads, the United States has fifty-seven hundred missile warheads to the Soviets' twenty-five hundred at the time the treaty is signed. However, the Soviets immediately push a MIRV program of their own.

The Antiballistic Missile Treaty limits both nations to deploying no more than two hundred ABM missiles, which would be designed to intercept attacking ballistic missiles. Both nations further agree not to develop a nationwide missile defense, but to concentrate their efforts on "point defense" of one specific site. The United States, deciding that existing ABM systems would not succeed against a determined missile attack, soon abandons its ABM program altogether. The Soviets continue with theirs. American intelligence concludes, at first, that the Soviet ABM system — which is built around Moscow — is aimed at defense against a relatively unsophisticated Chinese missile threat. But by 1983 it is rumored that American intelligence is worried by deployment of new Soviet ABM radar installations, which, if they are being built, would be a violation of the ABM Treaty.

As in so many attempts to control the arms race, the SALT I agreements are greatly weakened by a new acceleration of nuclear weapons' development, including the push to place MIRV warheads on missiles. More than half the American land-based missile force consists of the Minuteman IIIs, which carry three warheads. The Soviets begin replacing their older missiles with SS-17s, SS-18s, and SS-19s, which carry from four to ten MIRV warheads. The attempt to control the nuclear arms race has resulted in a new acceleration based on more sophisticated weaponry.

1974
Nixon resigns; Gerald R. Ford becomes U.S. President

1975
Viet Cong and North Vietnamese forces take Saigon; Vietnam united under Hanoi rule

Civil war breaks out in Lebanon

U.S., USSR, and 33 other nations sign Helsinki accord on human rights

1976
Jimmy Carter elected U.S. President

Syrian troops intervene in Lebanese civil war

1978
U.S. establishes full diplomatic relations with People's Republic of China

18 May 1974

India becomes the sixth member of the nuclear club by detonating a 10- to-15-kiloton atomic bomb in an underground test some hundred yards below the surface of the desert in Rajasthan, near India's western border with Pakistan. The underground explosion releases no radioactive debris into the atmosphere. The Indian government announces that it plans to produce nuclear explosives for peaceful purposes only, such as mining, oil and natural-gas prospecting, finding underground water, and diverting rivers; it has no intention of producing nuclear weapons.

Pakistan's Prime Minister Zulfikar Ali Bhutto, however, declares that if India develops nuclear weaponry, Pakistan's people "will eat leaves and grass, even go hungry, but we will have to get one of our own. We have no alternative."

1977

The Soviet Union begins deploying SS-20s in Eastern Europe. These medium-range missiles can reach any target in Europe, including the British Isles, even if they are launched from the Ural Mountains region of the Soviet Union. Western European governments ask the United States to upgrade NATO's missile capability, and the Carter administration responds by promising to deploy 108 Pershing II and 464 cruise missiles in Western Europe, beginning in late 1983.

18 June 1979

After more than seven years of dogged negotiations, President Jimmy Carter and General Secretary Leonid Brezhnev sign the SALT II agreement, which sets limits on the total number of nuclear "delivery systems" (that is, missiles and bombers) and on the number of individual warheads each missile can carry. But SALT II encounters heavy opposition in the U.S. Senate. As the debate over ratifying the treaty becomes heated, the Soviets invade Afghanistan, and Carter withdraws the treaty from consideration.

4 November 1980

Ronald Reagan is elected the fortieth President of the United States, partly on his promise to restore American military strength. In campaign speeches he stresses a "window of vulnerability" for American strategic nuclear forces: because the accuracy of the Soviets' missiles is improving, their superior nuclear forces could destroy

1979
Shah of Iran deposed; Islamic Republic proclaimed

Islamic revolutionaries seize U.S. embassy in Tehran, hold 63 Americans hostage for 444 days

Soviet troops invade Afghanistan

Camp David Accord signed by Egypt and Israel

much of the U.S. ground-based missile force in their silos, he claims, thus eroding America's strategic deterrent. He calls for building and deploying the MX missile, the B-1 bomber, and the Trident submarine-launched missile.

19 March 1981

The U.S. Department of Defense announces that the Soviet Union has orbited an "operational" antisatellite weapon, similar to the ASATs the Russians have been testing since 1968. General David C. Jones, chairman of the Joint Chiefs of Staff, testifies to Congress that "most of our military space systems [in low orbits] are vulnerable" to the Soviet ASAT.

7 June 1981

1981
Polish Army takes over government, declares martial law

Israeli planes bomb the Iraqi nuclear reactor facility in Osirak, near Baghdad. Prime Minister Menachem Begin says that the facility would have enabled Iraq to produce atomic bombs, and the pre-emptive strike was essential to prevent eventual nuclear attack on Israeli cities. The $275 million facility was being built by France and Italy. In response to world criticism, Begin says, "There won't be another Holocaust in history."

1981–1983

1982
Israel invades Lebanon

Responding to the planned and actual missile deployments in Europe

1982
Brezhnev dies, 10 November. Yuri V. Andropov named General Secretary of Communist Party, 12 November

1983
Andropov elected President of USSR, 16 July.

by NATO and the Soviet Union, and to Reagan administration statements about "battlefield missiles" and "war-winning capability," many Europeans begin to demonstrate against nuclear weaponry. These demonstrations grow into a massive Nuclear Freeze Movement, with the goal of stopping both East and West from deploying new missiles. The Nuclear Freeze Movement crosses the Atlantic and takes root in the United States, where many communities place nuclear freeze referenda on their local election ballots in November 1982.

23 March 1983
President Reagan proposes seeking a way to defend against nuclear attack, in a speech that includes his reasons for resisting an arms freeze and moving ahead with development of the MX. Aides later reveal that the President was referring to orbital ABM defenses that used lasers or other forms of energy-beam weapons. Critics claim the President is merely trying to get approval for his defense budget. The media dubs the idea "Star Wars."

3 May 1983
Roman Catholic bishops in the United States approve a stern pastoral letter opposing nuclear war and warning those who work in any phase

of nuclear weaponry that they are risking their souls.

The next day, the U.S. House of Representatives passes a nonbinding nuclear freeze resolution, calling on President Reagan to attempt to negotiate "a mutual and verifiable freeze and reductions in nuclear weapons" with the Soviet Union. The vote is 278–149. In October, however, the Senate rejects a similar nuclear freeze resolution.

August 1983

The U.S. Air Force delays the start of flight tests of a "direct ascent" ASAT system, a small rocket booster that is carried to high altitude under the wing of an F-15 fighter plane, then fired into space, where its warhead, a miniature homing vehicle (MHV), seeks out the target satellite and destroys it by direct impact. This system is cheaper and more flexible than the Russian ASAT, which requires a satellite-launching booster and needs several hours to establish itself in orbit and then match its orbit to its intended target. The delay is blamed on minor technical difficulties; tests are expected to begin by year's end or early 1984.

A few months earlier, a number of prominent scientists issue a petition to the United States government, urging the United States, the Soviet Union, "and other spacefaring nations to negotiate, for their benefit and for the benefit of the human species, a treaty to ban weapons of any kind from space and to prohibit damage to or destruction of satellites of any nation." Among the signers are Richard L. Garwin, of IBM's Watson Research Center; Carl Sagan, the widely known planetary astronomer; Hans Bethe, Lee Du Bridge, I. I. Rabi, George Rathjens, Victor Weisskopf, and Herbert York — all of whom have been intimately involved in the development of nuclear weaponry and strategic analyses.

1983
Soviet fighter planes shoot down Korean Air Lines flight 007 over China Sea, 1 September

October 1983
President Reagan and General Secretary Yuri Andropov continue to exchange ideas and insults on nuclear disarmament. Each side publicly rejects the other side's proposals while negotiators at the START (Strategic Arms Reduction Talks) and Intermediate Nuclear Forces (INF) conferences in Geneva work to find a common basis for counting strategic nuclear weapons, battlefield nuclear weapons, and the nuclear strike forces of Britain and France.

While Andropov proposes a treaty banning all weapons in space, the United States renews its promise to begin deploying Pershing II and cruise missiles in Western Europe, starting in December 1983. Massive public nuclear freeze demonstrations are expected. The United States announces that it expects the Soviet Union to walk out of the Intermediate Nuclear Force talks in Geneva as a protest against American deployment, over the next five years, of the 572 missiles in West Germany, Italy, and Britain. More than 200 Soviet SS-20s are already in place in the western Soviet Union, each carrying three independently targeted warheads, and new ones are being deployed at a rate of one per week.

17 October 1983

Colonel General Nikolai Chervov, a member of the Soviet general staff, is quoted by the weekly West German magazine *Der Stern* as saying that tactical nuclear missiles are deployed as a matter of routine "everywhere outside the USSR where Soviet Army divisions are stationed." He also warns that if NATO deploys the medium-range Pershing II and cruise missiles as planned, where the Pershings will be only ten minutes away from targets inside the Soviet Union, then Russia will establish missile bases within ten minutes' range of the continental United States.

18 October 1983

Defense Secretary Caspar W. Weinberger recommends to President Reagan the development of a space-based missile defense system, to be deployed before the end of the century. The recommendation is based on studies by panels of experts, headed by Defense Under Secretaries Richard De Lauer and Fred Iklé. The system is estimated to cost between $18 and $27 billion.

23 October 1983

Massive public demonstrations in Western Europe culminate a week of protests against the planned deployment of American medium-range nuclear missiles. More than 600,000 protesters fill the streets in Bonn, Hamburg, West Berlin, Stuttgart, and other West German cities. Hundreds of thousands jam London; an estimated 350,000 parade through the heart of Rome; and some 100,000 demonstrators "bring Vienna to a standstill," according to the *New York Times*. More thousands demonstrate in Paris, Stockholm, and Dublin. In the United States, thousands of peaceful protesters demonstrate in more than 140 locations; about 850 protesters are arrested across the United States.

25 October 1983

U.S. Marines and Army Rangers, together with token forces from other eastern Caribbean nations, seize the island nation of Grenada. President Reagan claims that Grenada was being turned into a Cuban-Soviet staging base for further subversion of other Caribbean and Central American nations. The action adds fuel to anti-American protests in Europe and Asia, where demonstrators depict the United States as a trigger-happy menace to world peace.

28 October 1983

NATO defense ministers decide to scrap fourteen hundred old nuclear warheads, and call on the Soviet Union to make similar reductions in its tactical nuclear arms. U.S. Defense Secretary Weinberger says that, although Peace Movement activists express alarm over the buildup of Western nuclear arms, "actually the trend is downward." Reaffirming their determination to deploy 572 new American medium-range missiles in Western Europe, the NATO ministers' decision to remove the 1400 older weapons — mainly artillery shells, antiaircraft missiles, and bombs designed to be dropped by airplanes — is characterized by Weinberger as an attempt by NATO "to maintain the [nuclear] deterrent . . . at the lowest possible level."

Soviet leader Andropov offers to destroy a hundred of the SS-20 missiles already emplaced in the European sector of the USSR if the United States agrees not to deploy its 572 Pershing II and cruise missiles. Andropov points out that the remaining 140 SS-20s based in Europe, with their three warheads each, would roughly match the number of nuclear missile warheads possessed by Britain and France.

16 November 1983

French President François Mitterand, speaking on television, says in regard to the growing Euromissile debate that "the East has developed missiles and the West has developed pacifists." Citing the Euromissile crisis as the gravest that has affected Europe since the Berlin blockade of 1948 and the Cuban missile crisis of 1962, the Socialist President of France insists that NATO must deploy the new American missiles or risk such an imbalance with the Soviet Union that "war is at our door."

24 November 1983

As the first American cruise and Pershing II missiles begin arriving in England, West Germany, and Italy, the Soviet delegation breaks off its talks with the United States at the

Intermediate Nuclear Force conference in Geneva, which had met for 105 sessions since opening, 1 December 1981. Soviet leader Andropov is quoted by Tass as saying, "The approximate balance of military forces, including nuclear ones, between the states of the North Atlantic Treaty Alliance and the Warsaw Pact states . . . has . . . served the cause of European security and stability. Now the United States and NATO as a whole are tipping the scales in their favor."

The following day Andropov announces that the Soviet Union will deploy more submarine-based missiles against the United States and step up its deployment of SS-20s in the western reaches of Russia.

But on 29 November West German Chancelor Helmut Kohl reveals that he has received a letter, signed by Andropov, in which the Soviet leader says that the impasse over the European missile situation is "not irreversible."

Writing in the *New York Times,* Mark Helprin points out:

> Both before and after the full American Euromissile deployments, the Soviet Union will have enjoyed and will continue to enjoy an often decisive quantitative advantage over all combinations of its opponents — in delivery vehicles, warheads and throw weight. . . .

> Despite the several millions who recently took to the streets in northern Europe, the "hot autumn" [of antinuclear demonstrations] was a good deal less hot than it was supposed to have been. . . . The governments of Britain, Germany and Italy — the three principals in deploying the missiles — won election on the platform of deployment, and in all cases . . . won it decisively.

30 November 1983

President Reagan gives his approval to a major program aimed at developing space-based missile defenses. Based on the Defense Department studies undertaken over the summer and autumn, the Star Wars program is expected to cost $18 to $27 billion during its first six years. The program is focused on energy-beam weaponry that would not be deployed until late in the 1990s or the first years of the twenty-first century.

8 December 1983

The fifth round of the American-Soviet strategic arms reduction talks (START) ends, after a thirty-minute session, with the Russians refusing to set a date for the next round. The negotiations on limiting or reducing nuclear arms began in 1969, were broken off, and then resumed in June 1982. Five rounds of two months each were held in Geneva.

The Soviet negotiator, Victor P.

Karpov, issues a statement linking the Euromissile situation to the bilateral U.S.–USSR talks: "A change in the overall strategic situation due to the beginning of the deployment of new American missiles in Europe compels the Soviet side to re-examine all the issues which are the subject of the discussion at the talks on the limitation and reduction of strategic armaments."

American observers point out that the Russian refusal to set a date for the resumption of START is not as severe as an absolute walkout. President Reagan comments that the Russians were "pretty careful with their words. . . . I think this is more encouraging than a walkout and simply saying they won't be back." Most observers do not expect a resumption of any serious arms negotiations, however, until after the American presidential election in November 1984.

9 December 1983

In a statement regarded by political observers as an "important political signal," NATO's foreign ministers offer the Soviet Union and the Warsaw Pact nations an opportunity to create a new, long-term relationship based on "moderation" and "an open, comprehensive political dialogue, as well as cooperation based on mutual advantage."

16 December 1983
The ten-year-long talks in Vienna on reducing conventional armed forces in Europe are suspended as the Warsaw Pact delegation, at the conclusion of the 358th session, refuses to set a date for their resumption. The talks, which had begun in 1973, are aimed at achieving a balance between Europe-based military forces of NATO and the Warsaw Pact. The level is to be lower than that which now exists.

Four of the seven Warsaw Pact nations are represented in the talks: the Soviet Union, Poland, Czechoslovakia, and East Germany. The East German delegate, André Wieland, cites NATO's deployment of American Pershing II and cruise missiles as the reason for the talks' suspension.

Thus, the only arms reduction discussions between East and West remaining active is the Conference on European Disarmament, which reopens in Stockholm on 17 January 1984, with thirty-five nations in attendance.

18 January 1984
Two days after President Reagan urged the Soviet leadership to join the United States in creating "a constructive working relationship" to solve East-West differences, Secretary of State George Shultz meets Soviet

Foreign Minister Andrei Gromyko at the foreign ministers' meeting in Stockholm for an intense five-hour discussion. In a speech earlier in the day Gromyko bitterly denounced the Reagan administration for "thinking in terms of war and acting accordingly. At present, the aggressive foreign policy of the United States is the main threat to peace."

Despite the public rhetoric, Shultz reports that the unexpectedly long meeting with Gromyko was "worthwhile. . . . Nobody was talking for effect this afternoon. They were professional diplomats approaching a problem seriously." But no problems are resolved, and no future meeting dates are announced.

1984
Yuri Andropov dies, 9 February; Konstantin U. Chernenko succeeds him as General Secretary of the Communist Party

21 January 1984
The U.S. Air Force launches its first ASAT weapon from an F-15 fighter plane in the beginning of a series of test flights off the coast of Southern California. No target was involved in the test, and no part of the two-stage rocket or its miniature homing vehicle warhead went into orbit.

2 March 1984
Konstantin U. Chernenko, the new Soviet leader, in his first public speech since the death of Yuri Andropov, urges the United States to take actions that would mark a turning point in

Soviet-American relations. "The Americans created obstacles to the talks both on European and strategic nuclear weapons by deploying their missiles in Europe," says the new General Secretary of the Communist Party. "It is the removal of these obstacles, which would also remove the need for our measures taken in response, that offers the way to working out a mutually acceptable accord."

Chernenko also says that "the U.S.A. can also make a not small contribution to strengthening the peace by concluding an agreement on the renunciation of the militarization of space."

On 12 March Chernenko tells the West German opposition leader, Hans-Jochen Vogel, that East-West talks can be resumed "at any moment" if the United States removes "the obstacles" it created by deploying the new missiles in Western Europe.

15 March 1984

Richard N. Perle, Assistant Secretary of Defense for International Security Policy, tells the Senate Armed Services Committee that a treaty banning antisatellite weapons would be "extremely difficult, if not impossible" to verify.

Writing in the April 1984 issue of *Aerospace America*,* Colin Gray, the

---

* Monthly journal of the American Institute for Aeronautics and Astronautics (AIAA).

president of the National Institute for Public Policy, says of antisatellite weaponry:

> The stable door is open, and the horse has long since bolted. Space weaponization is already unstoppable. . . . I do not suggest that ASAT arms control is unimportant, but rather that it would harm U.S. security. The Soviets know this, which is why they promote such a treaty. The overlap between ASAT and ballistic missile defense, the ease of deploying ASATs secretly, and the enormous incentive for the Soviets to cheat all make an ASAT treaty of no practical interest to the U.S.

16 March 1984

Representatives of NATO and the Warsaw Pact nations resume their negotiations of troop reductions in Vienna, suspended since December 1983. Diplomats from both sides caution against expecting a breakthrough in the ten-year talks.

Meanwhile, the Stockholm Conference on European Disarmament recesses for two months, following a strongly worded attack by the Soviet Union on NATO proposals for reducing the risks of a European war.

21 March 1984

The Union of Concerned Scientists issues a 106-page report, claiming that a space-based defense that would protect the civilian population of the

United States against nuclear missile attack is "unattainable," although a "modest" defense of specific facilities like missile silo fields is feasible. The report, based on a study conducted by Nobel laureate Hans Bethe, IBM scientist Richard L. Garwin, and seven other nuclear experts, warns that pursuit of a space-based defense system could lead to an escalation of the arms race and make nuclear war more likely.

27 March 1984

Lieutenant General James A. Abrahamson is named by Defense Secretary Weinberger to head the "strategic defense initiative" program to explore the feasibility of developing a space-based ABM system. Abrahamson, an Air Force general, had been directing the space shuttle program for NASA since 1981. The Star Wars program, funded at $1.98 billion for fiscal year 1985, is aimed at developing the technology necessary to make an "informed decision" about deploying orbital antimissile weapons in the 1990s.

5 April 1984

President Reagan announces that the United States will propose a treaty banning all chemical weapons, but that until such a treaty is in force, the United States must upgrade its chemical warfare weaponry. Since the proposed treaty includes provisions for on-site inspection of Soviet chemical warfare installations, experienced observers do not expect it to be acceptable to the Soviet Union.

Reagan's announcement is seen as an attempt to open negotiations with the new Soviet leadership on at least one aspect of disarmament while at the same time allowing the United States to catch up to the perceived Soviet lead in chemical weapons.

The same day, the Soviet leader, Chernenko, says that the United States has yet to make a concrete gesture toward reopening arms limitation talks with Russia. "Yes, a dialogue is necessary," says Chernenko. "However, the United States, though professing a desire for a dialogue, has not been backing it up with anything specific. . . . With each United States nuclear missile deployed on European soil, a new step is taken toward the danger line."

The history of the past forty years of wars, threats of wars, and excruciating international tensions has been largely a history in which the technology of armaments has continually outpaced political efforts to find a means to halt the nuclear arms race. A stability of sorts has been maintained, but only at the cost of building and fielding constantly more dangerous weapons: atomic bombs have grown to hydrogen bombs; the B-29 has evolved into the ICBM; surveillance satellites that monitor arms agreements are now threatened by ASAT weapons. The balance of nuclear terror has kept the superpowers from a major war, but each escalation in the destructive power of their weaponry reduces the margin of safety that keeps us from annihilation and extinction.

As Eugene Rostow, former director of the Arms Control and Disarmament Agency, put it, "The criteria for deterrence cannot be static, given the inevitability of endless change in technology."

Each attempt to negotiate an end to the arms race has only exacerbated the arms race. Nuclear missiles are sprouting in Europe and Asia, as well as in the fields of silos deep within the United States and the Soviet Union. The oceans are aprowl with missile-carrying submarines, and supersonic bombers bearing nuclear death in their bellies wing through the skies.

For four decades the emphasis has been on developing more powerful, more destructive weapons of attack. Offensive weaponry has outstripped all attempts at defense, as well as attempts to limit such arms through negotiations and international agreements. If the nuclear arms race cannot be stopped by good will, by common sense, or by hard-headed negotiations, can it be stopped or rendered ineffective by new types of weapons, based in space? Have we come to a point in history where defensive technologies can begin to overtake the runaway offense? Or will the

Star Wars concept merely extend the arms race into orbit and trigger the ultimate nuclear holocaust? The chapters that follow will examine these possibilities.

Tables 1 and 2 show the strategic nuclear weapons of the two superpowers as of early 1984 and the medium-range nuclear weapons that are being deployed in Europe. Taken together, these tables show the major nuclear attack forces fielded by East and West. They do not include the small but growing arsenal of the People's Republic of China or the nuclear weapons that other nations may be stockpiling in secret.

Table 1 shows the American and Soviet strategic arsenals in the operational inventories of the two superpowers as of December 1983. Note that the table shows strategic weapons *only;* tactical or battlefield weapons (also known as theater nuclear forces), such as the cruise missile or the SS-20, are given in Table 2, a tabulation of medium-range nuclear weapons and delivery systems presently deployed or scheduled for deployment in Europe. Table 2 does not include nuclear artillery shells, mines, and other short-range nuclear weapons.

Table 1: STRATEGIC WEAPONS

Missile	No. of Warheads	Warhead Yield*	No. of Missiles	Range (Miles)	Year Deployed
United States Land-Based Missiles					
Minuteman III	3	170–335KT	550	8000	1970
Minuteman II	1	*1–2MT*	450	8000	1966
Titan II	1	*9MT*	45	9300	1962
United States Submarine-Based Missiles					
Trident C4	8	100KT	264	4600	1980
Poseidon C3	10	50KT	304	2800	1971

United States Totals: 1613 missiles, 7297 warheads

		Soviet Union Land-Based Missiles			
SS-11	1	*1MT*	550	6500	1966
SS-13	1	750KT	60	6200	1968
SS-17 Mod 1	4	750KT	150	6200	1975
SS-18 Mod 2	8	900KT	107	6800	1977
SS-17 Mod 2	1	*6MT*	20	6800	1977
SS-18 Mod 3	1	*20MT*	26	6500	1979
SS-19 Mod 2	1	*5MT*	few	6200	1979
SS-18 Mod 4	10	500KT	175	5500	1979
SS-17 Mod 3	4	*2MT*	?	6200	1982
SS-18 Mod 4	10	500KT	?	6800	1982
SS-19 Mod 3	6	550KT	330	6200	1982

* KT means 1000 tons of TNT equivalent; MT means 1,000,000 tons TNT equivalent; warheads of megatonnage yield are in italics.

Table 1 continued

Missile	No. of Warheads	Warhead Yield*	No. of Missiles	Range (Miles)	Year Deployed
Soviet Union Submarine-Based Missiles					
SS-N-5	1	1MT	48	870	1964
SS-N-6 Mod 1	1	1MT	} 384	1500	1968
SS-N-6 Mod 2	1	1MT		1850	1973
SS-N-6 Mod 3	2	200KT		1850	1974
SS-N-8 Mod 1	1	1MT	} 292	4800	1972
SS-N-8 Mod 2	1	750KT		5650	1973
SS-N-18 Mod 1	3	200KT	} 224	4000	1977
SS-N-18 Mod 2	1	450KT		5000	1978
SS-N-18 Mod 3	7	200KT		4000	?
SS-N-17	1	1MT	12	2400	1977

Soviet Union Totals: 2378+ missiles, 6802+ warheads

United States Long-Range Manned Bombers: 272

Soviet Union Long-Range Manned Bombers: 143

Sources: James F. Dunnigan, *How to Make War* (New York: Morrow, 1982); *New York Times*, 12 and 24 April, 8 June 1984; International Institute for Strategic Studies, 1983.

Table 2: EUROMISSILES

System	Number Deployed	Warheads	Range (Miles)	Comments
NATO				
Pershing IA	180	1 each	450	Being phased out
Pershing II	108	1 each	1100	Deployment began December 1983; to continue through 1988
Cruise (GLCM)	464	1 each	1400	Deployment schedule same as Pershing II
Attack aircraft	1380	1 or 2 each	500	
SLBM	3 subs	400 total	2500	
Great Britain				
SLBM	64	192	2900	In 4 nuclear submarines
France				
Land-based missiles	18	1 each	1900	
SLBM	80	1 each	1600	In 5 nuclear submarines
Aircraft	135	1 or 2 each	300–500	
Soviet Union				
SS-20	243	3 each	3100	Numbers given are numbers deployed in Europe only; more than 100 SS-20s are deployed in Soviet Asia
SS-4	232	1 each	1250	Being phased out
SS-5	16	1 each	2500	Being phased out
Attack aircraft	3485	1 or 2 each	300	
SLBM	6 subs	18 total	900	

Sources: Congressional Research Service; *Newsweek* magazine, 24 October 1983; James F. Dunnigan, *How to Make War* (New York: Morrow, 1982); *New York Times*, 8 May, 13 October, 15 November 1983.

5

Why Fortify Heaven?

IN A WORLD that spends some $600 billion per year on armies and navies and air forces while half the planet's people suffer from malnutrition or outright starvation, is it really necessary to place weapons in space?

The average American firmly believes that space ought to be free of weaponry of all kinds; this new domain beyond the earth should not be turned into a battlefield. The average Russian undoubtedly feels the same way, although there are no public opinion surveys in the Soviet Union — at least, none conducted by independent, nongovernmental organizations. The late Leonid Brezhnev, General Secretary of the Communist Party and President of the Soviet Union, said in 1981, "May the shoreless cosmic ocean be pure and free of weapons of any kind." His successor, Yuri V. Andropov, suggested several times in 1983 that all antisatellite weapons be banned from space. However, Brezhnev's statement was made just as the United States space shuttle *Columbia* made its first flight; the Soviets have branded the shuttle a military weapon. And Andropov's offer came on the eve of the first test of the American ASAT weapon; the USSR's own ASAT has been an "operational" weapon for more than five years.

Keeping the infinite domain of space free of weapons and war is not as simple as it sounds. Although most Americans seem to feel instinctively that space is a new, clean environment that should not be polluted by human greed and aggression, the fact is that the military was operating in space long before anyone else. If it had not been for the military usefulness of space, neither the Soviet Union nor the United States would have started a space program. Governments do not risk billions of dollars (or rubles) for pure scientific knowledge or the romance of adventure. But they will spend significant portions of their gross national product to gain military advantage over their enemies — real or suspected.

It was the frightening power of the atomic bomb and, later, the still more awesome hydrogen bomb that made the military move into space all but inevitable.

Russian physicists knew just as much about nuclear physics as their German, British, or American counterparts did prior to World War II. There is evidence to suggest that the Soviet Union maintained a relatively small research effort in nuclear fission all through the war. And although the American Manhattan Project to develop an atomic bomb was carried on in the greatest secrecy possible, Soviet spies had penetrated the program and were sending information regularly back to Moscow. When, in July 1945 Truman hinted to Stalin that the United States had developed a new and very destructive weapon, Stalin apparently knew exactly what the President was talking about. He ordered a speed-up of the Russian atomic bomb work that very night.

Four years later the Soviet Union tested its first atomic bomb, far earlier than expected by most "experts" in the West. In 1951, after years of agonizing debate about the morality of developing the hydrogen bomb, which would increase the destructive power of nuclear weaponry a thousandfold, the United States conducted the Greenhouse George test, a "thermonuclear experiment" to see whether the hydrogen-fusion process really worked. It did, as was proved the following year by the Ivy Mike test, which wiped out the coral islet of Elugelab in a ten-megaton explosion. Later intelligence reports indicated that the Russians had also conducted a similar test several months *before* Ivy Mike. On 12 August 1953 the Soviet Union tested the first actual hydrogen bomb. Six months later, the United States tested its first deliverable thermonuclear bomb. Bernard J. O'Keefe, in his book *Nuclear Hostages,* pointed out, "The true father of the H-bomb is not Edward Teller, but Andrei Sakharov of the Soviet Union." Today, the sixty-three-year-old Sakharov, winner of the 1975 Nobel Peace Prize, is banished to "internal exile" in Gorki because of his antiwar and human rights activism.

A bomb, even one capable of demolishing an entire city, is not a weapon unless it can be delivered to its target. It was the quest for delivery vehicles to carry nuclear bombs that led the military into space.

The first vehicle to penetrate the virginal domain of outer space was Nazi Germany's Vergeltungswaffen Zwei, Vengeance Weapon Two, or V-2. This 46-foot-tall, 13.5-ton rocket-driven missile soared some sixty to seventy miles above the ground while traveling its two-hundred-mile course to deposit a ton of explosives on its target. The V-2's accuracy was abysmal: it could be used only as a terror weapon against very wide targets, like London.

But the combination of even a very inaccurate missile with an atomic bomb for a warhead was too obvious for military planners to ignore. In the United States, however, shortsightedness kept the American military from pushing toward rocket vehicles big enough and accurate enough to threaten cities thousands of miles away. Soon after World War II ended, the U.S. government asked Dr. Vannevar Bush to assess the possibility of building nuclear-armed missiles capable of intercontinental range. Bush had been director of the Office of Scientific Research and Development during the war, overseeing virtually the entire American R and D effort.

Nuclear bombs, at that time, were heavy, cumbersome devices that needed very large airplanes to carry them. The Little Boy bomb that destroyed Hiroshima weighed six tons; the Fat Man bomb dropped on Nagasaki was somewhat heavier and much wider, because of its different detonation system. Bush concluded that nuclear bombs were too big to be carried by rockets, and that rockets were too inaccurate to be used as long-range bombardment vehicles. The Air Force agreed, and pushed the development of manned bombers, such as the giant B-36 and, later, the B-52. In the late 1940s Bush wrote, "There has been a great deal said about a three-thousand-mile rocket. In my opinion such a thing is impossible for many years. I think we can leave that out of our thinking."

The Russians looked at the same evidence, though, and came to the conclusion that they would have to build very large rocket-driven guided missiles. By 1947, rocket development was virtually halted in the United States, while the Russians were beginning a development program that would see the first intercontinental missiles flying over five-thousand-mile test ranges by the mid-1950s.

The joker in the game was that scientists on both sides soon learned how to make nuclear weapons smaller and more compact, without sacrificing explosive yield, so that missiles could easily carry warheads of many megatons' explosive power. And missile accuracy has consistently improved to the point where the latest ICBMs, such as the Soviet Union's SS-18 and SS-19, can deposit their warheads within three hundred yards of ground zero. The American MX will reportedly have an accuracy of one hundred yards.

In the early 1950s American intelligence sources, primarily radar stations along the USSR's southern borders, detected the first Soviet missile tests. Suddenly Washington was in a sweat. If the Soviets perfected missiles that could drop hydrogen bombs on American cities, missiles that moved so swiftly that there was no way to stop them and scarcely any warning time for the threatened population, they could dictate terms to the United States and end the Cold War with a resounding Soviet victory — unless the United States had equally powerful missiles with which to threaten Soviet cities. Thus began the secret ICBM race of the 1950s, in which the U.S. Air Force pushed development of ICBMs while the Navy developed missile-carrying submarines.

In the midst of this race, Nikita Khrushchev decided to use Soviet missiles to enhance the USSR's prestige and political clout around the world. Sputnik I roared into orbit on 4 October 1957, the world's first artificial satellite, while the American satellite program, Vanguard, was not yet off the ground. I was a member of the Vanguard team, and even though we knew that the Soviets had more powerful boosters than we did, we were just as shocked and chagrined at their "first" as anyone. The pain and embarrassment were even worse when our first attempt to launch a Vanguard satellite exploded four feet above the launch stand, on 6 December 1957.

John Kennedy warned of the missile gap between the Soviet Union and the United States during his campaign for the presidency in 1960. After he gained the White House, however, American intelligence discovered that, though the Soviets had developed their missiles before the Americans had, they had failed miserably to produce large numbers of operational missiles. The United States, rushing to catch up to the Russians, had actually pushed far ahead of them. This was a major reason that Khrushchev was forced to back down during the Cuban missile crisis of 1962; the United States outgunned Soviet Russia, and the Kremlin knew it.

Once satellites started orbiting the earth, military thinkers realized that there were many things these artificial moons could do for them.

In a sense, the evolution of military activities in space has followed the pattern set by military air power. If ballistic missiles can be compared with artillery shells, then satellites can be compared with the earliest balloons and airplanes.

During the American Civil War and the Franco-Prussian War of 1870, hot-air balloons were sent aloft with observers dangling under them to look over the battlefield and "spot" for artillery; that is, identify targets and also report where the shells were actually falling, so that the gunners could adjust their aim and make the bombardment more accurate.

When the Wright brothers first showed their airplane to the U.S. Army, it was the Signal Corps that took an interest, in the hope that airplanes might make better artillery spotters. The Navy, too, was interested in airplanes primarily as platforms for gunnery spotting.

When World War I quickly bogged down into trench warfare on the Western front, airplanes began to be used for scouting the enemy's movements, miles behind the fighting front. One of my coworkers on the Vanguard project was an ex-Canadian who had flown a night bomber in World War I. He was a colorful, crusty old man, full of stories about what it was like to fly in those days of open cockpits and rickety, fabric-covered aircraft. He claimed to have invented the idea of armor plating for airplanes: a bullet from the ground once came through his seat, between his legs, so on his next flight he protected himself by placing a stove lid on the seat.

In the early months of the war, he told me, there was no air-to-air fighting. "It was dangerous enough up there, without shooting at people," he said. But slowly, military men began to realize that the airplane could be used not only to observe the enemy, but to do damage to him. Bombs could be dropped on enemy troops, on supply depots and ammunition dumps, on railroad trains and truck convoys, even on the rail tracks and roads. Troops could be strafed with machine guns from the air, even though they might be miles behind the fighting front.

In fact, the "fighting front" could be extended even into the enemy's heartland from the air. When Louis Blériot made the first flight from France across the English Channel, on 25 July 1909, a shudder went through knowledgeable Britons. They realized that their island was no longer completely defended by the Channel from attack and invasion.

Less than six years later, German airships began to bomb England. On the night of 19/20 January 1915, two Zeppelin dirigibles bombed the port city of Yarmouth. Their targets were the docks and shipyards, but the night bombing was inaccurate; bombs damaged the town square and demolished several houses. Four people were killed. At first only the giant Zeppelins had the range to reach England; later in the war German Gotha bombers joined the attack. But the Zeppelins carried most of the burden. Until almost the last months of the war, these lighter-than-air craft droned through the night skies to bring the war to civilian men, women, and children. Their attacks were sporadic, limited by weather and moonlight conditions. The British defended themselves with interceptor airplanes and antiaircraft guns. The hydrogen-filled Zeppelins were huge targets that burst into flame when hit — funeral pyres floating through the night skies.

The Zeppelin raids did little to affect the outcome of the war. But they brought home the gloomy predictions of men like H. G. Wells, whose science fiction tales years earlier had foretold of vast aerial bombardments from which no one would be safe.

As aircraft — Zeppelins and winged airplanes — became formidable in their attacks on ground targets, it was only natural that other aircraft should be used to intercept and destroy the attackers. Thus the cruel evolutionary pressures of war forced the rapid development of fast, nimble fighter planes with reliable engines and rapid-firing machine guns. A new factor entered warfare: air superiority. If you could maintain superiority in the air over your enemy, you could attack his forces on the ground while preventing his planes from attacking your ground troops. Control of the air was not crucial in World War I. It became so in World War II.

Between the two wars, prophets of air power like Italy's Giulio Douhet, Britain's Hugh Trenchard, and America's "Billy" Mitchell created the ideas and techniques that would result in the devastation of cities across Europe and Asia. They predicted that future wars could be won from the air by "strategic" bombing of the enemy's industries and cities, which would bring his war-making capacities to a halt without the need for battle between land armies.

Beginning with the Nazi terror bombings of Warsaw and Rotterdam, accelerating into the Blitz against London, and culminating in the Combined Bomber Offensive staged by the RAF and the U.S. Army Air Forces against Nazi Germany, close to two million tons of bombs were dropped by airplanes on civilian targets in Europe during World War II. Two megatons: less than the explosive yield carried by a single missile today.

The Japanese also received a bitter dose of aerial death, delivered by B-29s based in China and the Mariana Islands of the central Pacific. Tokyo, Osaka, and a dozen other Japanese cities were virtually wiped out by massive incendiary raids; 125,000 civilians were burned to death in a single fire raid on Tokyo the night of 9/10 March 1945, the victims of two-pound incendiary bombs carried by 334 B-29s. Some four million pounds of thermite and oil bombs razed sixteen square miles of the Japanese capital, destroying one fourth of the city's buildings. Osaka, Kobe, and Nagoya were similarly gutted; the air raids stopped only when the Americans ran out of incendiary bombs and had to wait for replacements. Within three weeks they were back over Tokyo again, destroying another eleven square miles and killing thousands more in firestorms that raged out of control.

In August the final horror of Hiroshima and Nagasaki broke the Japanese will to fight. The war ended without the need to invade Japan's home islands, an invasion that would have caused millions of casualties on both sides. The atomic bomb — and those two-pound incendiaries — also provided the ultimate vindication for Douhet and other early prophets of strategic air power. A modern military state had been defeated without recourse to invasion and ground battle, defeated largely by strategic bombing. Of course, as naval experts correctly pointed out, the Japanese home islands had been strangled by a U.S. Navy blockade and submarine offensive that had cut off virtually all the petroleum and other vital raw materials Japan had to import from Southeast Asia.

Japan was starving to death in 1945. Dr. Taro Takemi, a former president of the Japan Medical Association, said in 1983 that "many people would have starved if the atom bomb had not been dropped," forcing the Japanese military to admit defeat and surrender. Dr. Takemi, writing in the *Journal of the American Medical Association* on the thirty-eighth anniversary of the Hiroshima bombing, said, "When one considers the possibility that the Japanese military would have sacrificed the entire nation if it were not for the atomic bomb attack, then this bomb might be described as having saved Japan."

The atomic bomb made triumphant the concept of strategic bombing. Attacking from the air the enemy's heartland cities, factories, oil refineries, railroad centers, shipyards — and civilian workers — became the cornerstone of strategic thinking. The Air Force, the "glamour boys" of World War II, became the primary striking arm of the modern military services. The Navy and Army shrank in prestige, size, and appropriations. General George S. Patton growled that the atomic bomb allowed "pacifists, politicians and fools . . . to say, 'All we need is a bomb, no Army.' " And, indeed, the size of the U.S. Army was reduced from more than eight million men to 690,000 between 1945 and 1947.

Within ten years, American foreign policy as enunciated by President Eisenhower's Secretary of State, John Foster Dulles, was based on the concept of *massive retaliation* against any aggressive move made by an enemy. Instead of mobilizing troops or dispatching a fleet of warships, we would drop nuclear bombs on the nation that threatened us.

The means for delivering those weapons, by the year 1960, was the ICBM, the "ultimate weapon," a missile that could not be intercepted, could not even be recalled once it was launched, and that carried more explosive power in its warhead than all the bombers in all the missions of death flown throughout World War II.

The ICBM spends most of its flight time in space, miles above the fringes of the earth's atmosphere. By 1957, warfare had truly entered space.

6

Naked to Mine Enemies

WHEN THE RUSSIANS sent Sputnik I into orbit, American media reacted with shock and alarm, calling the event "a technological Pearl Harbor."

That phrase was more than dramatic rhetoric. But perhaps it would have been more accurate to say that Sputnik represented not so much an attack, as Pearl Harbor was, but a warning, an omen of the impending possibility of attack. Just as Blériot's flight across the English Channel shattered Britain's feeling of security, Sputnik made many Americans realize that the oceans separating the United States from Europe and Asia were no longer much protection. Rocket-boosted hydrogen bombs could reach American cities within a half-hour of being launched. And nothing on earth could stop them.

There have been arguments in the United States and other democracies over the legitimate role of the military services. There have been equally bitter arguments over the possibilities of extending military activities into orbital space. But if it can be agreed that the minimal role for the military services is to protect the nation's borders from attack, it must be realized that *we have a border with space.* Every square inch of this nation, of every nation, is open to attack from space.

Like the evolution of aerial warfare, space warfare is moving swiftly toward more and more destructive capabilities. First came the missiles, the ICBMs and SLBMs that arc high above the atmosphere and carry their hydrogen-bomb warheads thousands of miles into the heartland of the nation under attack.

As with the airplane, the first use that military planners saw for satellites was as observation platforms and very high altitude "artillery spotters" — the "artillery" being intercontinental ballistic missiles. But as the ICBMs were being deployed, American strategists began to realize that the maps they had on hand to designate the targets for their missiles were not accurate enough to be usable.

Buckminster Fuller, of geodesic-dome fame, told me once that he had stunned a group of Pentagon officers by proving to them that their maps of the Soviet Union were so inaccurate that missiles would miss most of their targets by miles. While the United States and Western nations in general have such open societies that precisely detailed maps can be bought by anyone, such maps of the USSR are not available, even inside the Soviet Union and to its own citizens.

Thus, one of the most critical uses of artificial satellites became *geodesy*, the study of the size and shape of the earth. Using the findings of the satellites, cartographers are able to make maps as precisely accurate as the guidance systems of the missiles themselves. Although the Vanguard satellite program had its problems and very public failures, we did succeed in orbiting three satellites out of a total of nine launches. (Not a shabby record, actually, for a completely new rocket booster, in those days of the late 1950s.) The first Vanguard satellite, put into orbit on Saint Patrick's Day of 1958, with a Saint Christopher's medal welded to its guidance section, was used by geophysicists to make beautifully exact determinations of the earth's shape and size. Later geodetic satellites have been lofted precisely for the purpose of making extremely accurate maps.

Satellites are excellent platforms for making all kinds of observations. From the use of scouts perched in trees or atop hills to hot-air balloons to airplanes, the military has sought the means to peer as deeply as possible behind the enemy's front line. During World War II, "pathfinder" planes usually preceded major bomber formations to scout the target area and radio back reports on weather conditions over the target. The culmination of aircraft reconnaissance was the U-2, RB-57, and the Mach 3 SR-71: planes capable of flying well above eighty thousand feet for thousands of miles into unfriendly territory. But the world-rocking Francis Gary Powers incident, in which the Russians shot down an American U-2 a few days before an Eisenhower-Khrushchev summit conference was to begin in Paris, showed that antiaircraft missiles could reach even the highest-flying airplanes.

Military reconnaissance moved from airplanes to satellites, although the U-2 and its descendants are still being used in many places around the globe. It was a U-2 camera that revealed that the Russians were building missile bases in Cuba in 1962. However, satellites offer several advantages over airplanes. They orbit well above the range of conventional antiaircraft defenses — although ASAT weapons are ending the era of satellite invulnerability. Satellites can remain in orbit indefinitely, and their cameras can keep watch on the earth below for as long as the satellite's equipment remains functional. American reconnaissance satellites remain functional for months, years. Soviet satellites, presumably because of their inferior electronics equipment, are deactivated and replaced by new satellites every few weeks.

The Soviet Union protested vehemently against American "spy satellites" when the first Discoverer observation satellites were orbited, in the early 1960s. But their protests were without much legal force, because it was Soviet Russia itself that had established the precedent of orbiting satellites over other nations without prior permission. When Sputnik I stunned the world, in 1957, it also sent the international legal fraternity into something of a tizzy. Since Roman times, property rights had been assumed to extend up into the air and on into infinity. When aircraft began flying across national borders, international agreements had to be worked out to permit such overflights. Every nation had, and still has, the legal right to deny another nation's aircraft access to its airspace. In the extreme, this right can be backed by force — just as the Russians did when they shot down a Korean Air Lines commercial jetliner in 1983 because it had strayed into Soviet airspace.

But Sputnik beeped away up in orbit, crossing the borders of every nation on earth without any permission from any government. No one could do anything about it, because there was no way to shoot down the satellite — even if any nation had wanted to take that extreme step. Legally, the Russians had established the de facto principle that air rights end at the upper fringes of the atmosphere. Satellites, in space, can orbit across national boundaries without hindrance.

The Russians did not like having American cameras peeking at them from orbit, but there was little they could do about it, legally. They soon enough began orbiting their own reconnaissance satellites, and then their protests ceased. But they also took another step: they began to develop ASAT weapons so that violations of their borders could be met with force, when and if the Kremlin decided to do so.

Because surveillance satellites orbit five to ten times higher than even the highest-flying airplanes, their cameras cannot pick out as much detail as the cameras on a plane. In technical jargon, this is a matter of *resolution:* the capability to distinguish small details in a picture. To obtain the best possible resolution, the satellite must be orbited at as low an altitude as possible. But a satellite flying below a hundred miles will encounter so much friction with the uppermost wisps of the atmosphere that it will plunge back to earth within a few orbits. So even the lowest-altitude satellites orbit above a hundred miles.

On the other hand, the lower the orbit, the faster the satellite is moving. Johannes Kepler discovered this law of planetary motion in 1609, and it applies just as much for an artificial satellite as it does for the planet Mars. Keeping a satellite in the lowest possible orbit means that it will sweep across the ground below it at the fastest possible speed, which is not always desirable for reconnaissance missions. Usually a compromise is struck, and the satellite is orbited at an altitude that will allow good photographic resolution and a reasonable "dwell time" over the areas of interest.

During the Falkland Islands War in 1982, for example, the Soviet Union launched several observation satellites into highly elliptical orbits. They reached their highest altitude (apogee) over the South Atlantic, which meant that they spent more time over that region of the earth than any other. Evidently the Soviets were willing to swap picture resolution for dwell time.

Of course, the quality of the camera plays a very strong role in the ultimate resolution. For years the story has been told, jokingly, in the halls of the Pentagon that our reconnaissance satellite photographs are so sharp that you can read the numbers on the license plates of the cars in the Kremlin's parking lot. Once in a while, someone will hint that you can read, from orbit, a newspaper over the shoulder of a Muscovite.

The first reconnaissance satellites, in the early 1960s, used film cameras. Once their film was completely used up, the satellite was programmed by ground controllers to eject the film cassette, which was protected by its own little re-entry vehicle. The cassette came down through the upper regions of the atmosphere and then popped a parachute, very much in the way the Mercury, Gemini, and Apollo astronauts returned to earth, except that the film cassette was snagged in midair — most times — by an Air Force cargo plane specially rigged for the task. As television cameras improved in quality, electronic video links began to replace the film cameras, allowing the satellites' useful lifetimes to be extended far beyond the limits of even the largest film canisters.

The original Discoverer of the early 1960s gave way to more sophisticated observation satellites, such as Samos and Midas. The Soviets began orbiting their own observation satellites as part of their Kosmos program. Kosmos is a catch-all name for a variety of scientific, communications, weather, and military satellites. More than twelve hundred satellites titled Kosmos have been launched since the mid-1960s. The reconnaissance types weigh more than six tons each, and the latest models are maneuverable: they can shift from a low orbit to a higher one, or vice versa, under remote control from the ground.

On 15 June 1971, the U.S. Air Force launched its first "Big Bird" surveillance satellite. Officially named Project 467, Big Bird weighed about three tons, was nearly fifty feet long, and had optical equipment good enough to allow it to be placed into a higher orbit than earlier observation satellites. Big Bird held five film re-entry capsules and had an operational lifetime of five months. On 19 December 1976, a newer type of satellite, KH-11, carried into orbit a set of television cameras that could relay information to the ground electronically.

Today, the United States employs several different types of reconnaissance satellites: the Big Birds appear to be the backbone of the system, with KH-11 satellites orbiting at somewhat higher altitudes and remaining useful for years rather than months. Smaller, "close-look" satellites are sent up into the lowest orbits when there is an area of particular interest to be surveyed.

Reconnaissance satellites are useful not only to intelligence analysis of the CIA and National Security Agency. They can be extremely important to tactical military commanders, offering them a view of battlefields and the enemy's rear elements that would be difficult or impossible to obtain with aircraft or other means. Nor are satellite sensors restricted to optical cameras. Infrared cameras, which sense heat waves, and radar can penetrate clouds and darkness. By combining these sensors with highly sophisticated computerized equipment on the ground, it is possible to produce pictures of astounding clarity and detail, even at night or in bad weather.

On the strategic side, of course, the satellites can watch a potential enemy's homeland, keep track of ships at sea, and even offer information on submerged submarines. Electronic "ferret" satellites can eavesdrop on radio and telephone conversations for intelligence purposes. For years now, American analysts have been able to make very detailed and accurate forecasts of the Russian grain harvests, because American satellites can measure the amount of grain ripening in the Ukraine and elsewhere.

Perhaps the most vital function of observation satellites is early warning of missile attack. From orbit, satellites can easily see the fiery plume of a rocket launch. Early-warning satellites stand guard against the moment when a hundred or a thousand rockets are launched simultaneously. They will not be able to stop those missiles from reaching their targets, but they will provide enough warning — in theory — to allow a counterstrike to be launched against the attacker. Thus is played the game of deterrence. If the attacker knows he will pay in full measure for his aggression, he will probably stay his hand. By providing enough warning time to allow us to launch our counterstrike, the early-warning satellites help to prevent the aggressor from launching an attack in the first place.

Certainly these early-warning satellites are vulnerable to ASAT weapons. If these satellites are suddenly and methodically destroyed, however, it would be tantamount to warning that an attack is either planned or already on the way. Would an American President order a nuclear strike if our early-warning satellites were incapacitated? Or would he hesitate and wait, his advance scouts blinded, until the attacker made his next move?

Although reconnaissance satellites have become crucially important to the military, they also have become central to efforts at disarmament. For more than a quarter of a century, scientists and diplomats on both sides of the Iron Curtain strove to find a way to reach some kind of agreement that would halt, or at least slow, the nuclear arms race between the United States and the Soviet Union. All attempts at an agreement foundered on one basic point: the United States would not sign any agreement that did not include inspection of the Soviet military establishment, and Russia would not allow any such "spying" on its territory.

Satellites broke the impasse. By the late 1960s it became apparent to both sides that the other nation was counting its missile silos and submarines. Such data were, in fact, a form of inspection, whether the nation being inspected liked it or not. This broke the logjam over arms limitation and permitted the SALT I agreements. Future arms limitation agreements, whether they are called SALT II or START or something else, will rest heavily on satellites that watch from orbit. The diplomats call such observation satellites and associated systems "national technical means for verification" of the arms agreements.

Another type of observation satellite stands watch over the Nuclear Test Ban Treaty. The treaty forbids the testing of nuclear weapons on the earth's surface, under water, or in space. In 1963 the United States launched its first pair of Vela Hotel satellites, tiny (105 pounds) watchdogs that were hurled into very high orbits, nearly sixty-seven thousand miles up, on opposite sides of the earth. Since then, Vela Hotel satellites and their descendants have patrolled the skies, on the lookout for clandestine nuclear tests on the earth's surface or in space. Scientific sensors riding piggyback aboard such satellites have discovered several new phenomena in space that have excited the astronomical community.

In 1979 a Vela Hotel satellite detected what appeared to be a nuclear detonation in the South Atlantic Ocean. Rumors flew across the world that it was a nuclear bomb test carried out by Israel or South Africa or by the two nations working in concert. After careful analysis of the data, though, the U.S. government announced that the satellite's detectors had picked up nothing more than the glint of sunlight off the ocean waters. Many observers of the international scene were unconvinced by that explanation and are certain that Israel not only has the know-how to produce nuclear weapons, but has actually tested one.

Observation satellites watch the weather, as well as military hardware and troop movements. We take it rather for granted now that satellite photographs of the earth's weather systems adorn our newspaper and television weather forecasts. But I remember when such ideas were science fiction.*

The military is just as vitally interested in accurate weather forecasts as civilians are. A drenching rain bogged down Napoleon's cannon in mud at the Battle of Waterloo. Japan's attack fleet cruised to Pearl Harbor under cover of a storm front that prevented detection from the air. The D-Day invasion of Normandy hinged crucially on an accurate forecast of the weather of 6 June 1944. Nazi Germany's final gasp, the desperate offensive known as the Battle of the Bulge, took place during snowy winter weather that grounded the Allies' overwhelming air forces.

Weather satellites are usually placed in *geosynchronous* orbits 22,300 miles above a spot on the equator, where they orbit the earth in the same twenty-four-hour period that our planet itself revolves. This keeps them stationary over that spot on the equator, and enables them to "see" almost half the planet all the time.

Like the reconnaissance satellites, weather satellites no longer depend on cameras alone. Infrared sensors can measure the heat emitted by the ground and the seas; infrared systems can now give almost as good a picture of the nighttime weather as normal optics give of the daylight situation. Other sensors can measure directly or allow meteorologists to deduce the temperature of the ground, water, or air being examined. New radars, tested within the past two years aboard the space shuttle, will allow weather satellites to measure the waves in midocean precisely enough to determine the force of the winds there.

The other major function that satellites perform for the military is one that they also perform for us civilians: com-

---

* In fact, in the mid-1960s I wrote a novel, *The Weathermakers,* that dealt with the subject.

munications. Today, roughly 80 percent of the U.S. armed services' communications around the world are relayed through satellites.* If those commsats were to be suddenly disabled, American forces overseas, ships in midocean, submarines on patrol, would all be abruptly cut off from their high commands.

Picture a Trident submarine, carrying enough destructive power in its missile warheads to obliterate a continent, suddenly unable to make contact with its fleet headquarters or with *any* other units of the Navy. What would go through the mind of that submarine's captain? What do his orders say about such a situation? Would he launch the sub's missiles? Would his orders forbid him to do so?

The communications satellite is the brainchild of Arthur C. Clarke, the famous author of *2001: A Space Odyssey* and hundreds of other fine works of fiction and nonfiction. In 1945, while he was working on radar systems for the British defense establishment, Clarke hit on the idea of using artificial satellites to relay communications. Essentially, he realized that the higher you can place a relay station, the wider an area it can cover. A satellite orbit struck Clarke as the highest "pole" conceivable for a communications relay. And the best orbit for a communications satellite would be the geosynchronous, twenty-four-hour orbit, where the satellite appears to hover in the sky, so that antennas on the ground can easily be locked onto it and do not need to track a moving relay station. That geostationary orbit is now known as the Clarke orbit.

Like weather satellites, communications satellites are so commonplace today that we tend to take them for granted. The military does not, because they cannot afford to. With

---

* The usual abbreviation for communications satellites is commsats, with two *m*'s. Comsat looks better, but it happens to be the registered trademark of the Communications Satellite Corporation, and Comsat Corp. takes a dim view of having its name used generically.

nearly 80 percent of its communications relayed around the world through commsats, the U.S. military must pay careful attention to its satellite network. When the 1962 Starfish high-altitude nuclear test knocked several orbiting satellites "off the air," scientists and military planners soon realized that the huge thump of radio-wave energy, the so-called electromagnetic pulse (EMP) that blasts out of the nuclear fireball with the speed of light, can disable the electronics systems in satellites over vast distances. A 1-megaton explosion above the atmosphere could knock out all the commsats on that side of the globe!

EMP can also raise havoc with ground communications, causing electrical surges in all sorts of electronic equipment — even home television receivers and underground cables. For more than twenty years the Defense Department has been working on "hardening" its communications systems on the ground and in satellites against EMP. Many of the underground nuclear tests conducted in the American Southwest have been aimed at EMP experiments.

Even if the Defense Department's commsats are hardened sufficiently to withstand EMP, civilian commsats are not. If someone detonated a nuclear bomb in space, the resulting EMP might knock out most of the commsats that now routinely relay long-distance telephone and television communications. Much of the United States' day-to-day commercial business depends on such cross-country and transoceanic communications. Knocking out America's civil commsats would cause havoc for American business.

EMP has added a new worry to civil defense: the Chaos Factor. A nuclear attack might begin with the explosion of a massive bomb at high altitude, above the heartland of America's Midwest, which would blast out an EMP that would knock out telephones, radios, computers, car ignitions, perhaps even trigger electrical surges in the control systems of nuclear power plants that could lead to shutdowns — or even meltdowns — of the nuclear reactors. One such bomb might so incapacitate the United States that there would be no need for further attack; the nation would be crippled already. Is that why the Soviets tested a 60-megaton bomb?

To Clarke, the father of the commsat, the idea of military operations in space is anathema. As we shall see in later chapters, Clarke has been one of the leaders in the attempt to ban all weaponry from space.

But the inexorable logic of military evolution dictates that active weapons will be placed in space, just as armed fighter planes took off after the scouting aircraft of World War I. Fearing that the Soviet Union would place nuclear bombs in orbit, the United States began testing a crude ASAT system in the 1960s: a small nuclear warhead lofted into space by a Thor missile. To their shock, the military found that even a small nuclear explosion in orbit fired out an EMP that ruined the electronics on satellites halfway around the world, no matter whose satellites they were. The EMP problem, and the Outer Space Treaty of 1967, put an end to that first ASAT system.

The Russians, though, soon began testing their ASAT, which bears a warhead of conventional explosive. One way to protect American satellites is to threaten Soviet satellites; retaliation is often easier than actual defense. So the U.S. Air Force is developing its own ASAT system.

Now, after more than a quarter-century of placing passive military systems in orbit, the superpowers can use active weaponry in space. Neither the Soviet ASATs nor the American ones are actually based in space: they wait on the ground until they are needed. But the next logical step in the militarization of space is the placing of ASAT weapons in orbit, where they can wait quietly, disguised perhaps as reconnaissance or scientific satellites, until their human commanders order them into action.

Washington rumor has it that ground-based lasers have already been used to attack satellites. On more than one occasion, according to such stories, American reconnaissance satellites have been blinded by laser beams fired from Soviet territory. The Pentagon has admitted that at least one U.S. reconnaissance satellite's optical system was damaged while the satellite was passing over Siberia, but laid the blame on an intense fire in a natural-gas field that overexposed the camera's film. Whichever explanation is the truth in that case, powerful lasers — based either on the ground or in space — would be able to demolish most satellites, which are not built to withstand intense beams of energy. In the parlance of the military, satellites are "soft targets."

The Arms Control and Disarmament Agency reported to President Reagan in late 1983 that U.S. Air Force Satellite Data System and British Skynet 2 satellites may have been consistently illuminated by Russian ground-based lasers when they passed over Sary-Shagan, the site of a Soviet laser test facility. SDS and Skynet 2 are communications satellites that have "suffered temporary anomalies while over the USSR," according to *Aviation Week* magazine.

Most scientists on both sides of the Iron Curtain have appealed to their governments to ban all weapons in space. The Soviet Union has officially submitted a draft treaty to the United Nations that calls for outlawing space weapons. But the treaty defines the U.S. space shuttle as a weapon. And the Pentagon is loath even to consider a ban on space weapons until the Air Force's ASAT system becomes operational. Otherwise, the Soviets will have an ASAT ready to be used whenever they want to use it, but the United States would have no similar capability.

The arms race in space, evidently, will follow the weary path of earlier arms races: neither side is willing to halt its arms developments unless it feels that it is "ahead" of the other side. Each side claims that it is trying to "catch up" to its opponent. In reality, each is seeking a way to attain superiority over the other. Only in superiority can the military feel secure. But the quest for such superiority guarantees that the arms race will continue.

Clearly, the military functions of satellites — reconnaissance, early warning, weather observation, nuclear test-monitoring, communications, and mapping — can play as important a role for peacekeeping as for war-making. The ability to see a nation's military panoply, to observe troop movements, construction projects, concentrations of armor, missile launches, all these are vital to any peacekeeping force. So is the ability to communicate such information from any spot on the globe (or in space) to the decision-makers who control the peacekeepers.

Because military satellites are already so important to the strategic and tactical leaders of the world, such satellites are very tempting targets. As we have seen, ASAT weapons have already been developed by both the superpowers; recent history has shown that such weaponry soon enough "trickles down" from the superpowers to their close allies, then to the more belligerent of the nonaligned nations. Some rather sophisticated weaponry, such as heat-seeking antiaircraft missiles, have even found their way into the hands of terrorists, who have used them on occasion to shoot down commercial airliners.

The evolution of space warfare is indeed following the general lines of the evolution of air warfare. Command of orbital space will be just as important in a conflict between the superpowers as command of the air has become for ground or sea battles. Satellites, like the early biplanes of World War I, are now used as scouts and artillery spotters. The "fighter planes" are not far behind — ASAT systems that can attack the other side's unarmed scouts.

As aerial warfare evolved in the two World Wars, fighter planes found themselves battling against the other side's fighter planes for command of the air. The next stage of the militarization of space may well find satellite "battle stations" in orbit, armed with hypervelocity missiles or energy-beam weapons, fighting the other side's "battle stations" for the command of space.

For whoever commands orbital space can see his enemy, from battlefront to heartland, can communicate what he sees instantly across the globe, can direct hydrogen bombs or other weapons onto his enemy's head without hindrance, and can — or will someday be able to — destroy his enemy's ballistic missiles long before their hydrogen-bomb warheads get close to their targets.

That position of dominance is far too important for one superpower to allow another superpower to achieve. Neither the United States nor the Soviet Union can afford to sit by and allow the other to take such command of orbital space. Either nation would go to war to prevent such a situation.

But would the superpowers be willing to allow an international agency, a nonaligned peacekeeping force, to attain command of orbital space? It may be the only way to avert the long-dreaded nuclear war between them.

7

"Star Wars" vs. "MAD"

IT BEGAN, as most things do, in a science fiction story.

> Suddenly there was a flash of light. . . . At the same time a faint hissing sound became audible. . . . Forthwith flashes of actual flame, a bright glare . . . sprang from the group of men. It was if some invisible jet impinged upon them and flashed into white flame. It was if each man were suddenly and momentarily turned to fire. . . . It was sweeping round swiftly and steadily, this flaming death, this invisible, inevitable sword of heat.

Thus H. G. Wells described the "heat ray" weapon used by the invading Martians against helpless Earth men in his novel *The War of the Worlds,* which was published in 1898.

Exactly four decades later, the actor-director Orson Welles dramatized *The War of the Worlds* in a Hallowe'en radio broadcast that terrified thousands of listeners with its realism and convinced them that Martians had actually landed in New Jersey and were destroying everything in their path. In the late 1960s I heard a recording of that broadcast, and was fascinated to listen to the sound that Welles had chosen for the Martians' "heat ray": a throbbing, high-pitched whine. For I had heard a very similar sound in the laboratory where I was working: the throbbing, high-pitched whine of a powerful electrical generator that was "driving" a high-power gas laser — the real-world equivalent of the Martian "sword of heat" that Wells had envisioned more than two generations earlier.

Ever since *The War of the Worlds*, evil monstrous creatures from other worlds and all-destroying "death rays" have been staples of science fiction. Through the 1930s and afterward, the lurid covers of science fiction magazines almost invariably pictured a BEM (bug-eyed monster) carrying off a scantily clad human female with one tentacle and waving a "ray gun" in the other. Death rays and disintegrator weapons were the standard sidearms for Buck Rogers, Flash Gordon, and a bevy of science fiction and comic strip heroes.

But give a science fiction idea enough time, and it often turns into everyday reality. In the 1940s another of Wells's predictions — the atomic bomb — blasted its way into the real world. By the mid-1950s the Soviet Union and the United States were galloping, hell-bent, in a race toward a double goal: nuclear weapons and ballistic rocket missiles. Still, if you mentioned energy-beam weapons to any respectable scientist or military officer, he would snort and say that "death rays" were about as likely as flying to the moon.

Yet in the 1950s the theoretical work leading to the invention of the laser was being done by Charles Townes of MIT, Arthur L. Schawlow of Bell Labs, and A. M. Prokhorov and N. G. Basov of the Lebedev Institute of Physics, in the Soviet Union. In 1960 Theodore Maiman, at the Hughes Research Laboratories in California, produced the first working laser. It was a tiny cylinder of artificial ruby less than two inches long; the pulses of pure red light it emitted lasted only half a thousandth of a second and were much less than 1 watt in power. But like the brief, wobbly flight of the first Wright brothers' airplane, that first laser was important not so much for its actual performance as for what that performance presaged for the future. Within six years, the scientists at Avco Everett Research Laboratory had invented the high-power gasdynamic laser.

By 1971, when I left the laboratory, Avco had sold a relatively low-power gasdynamic laser to the Caterpillar Tractor Company. Caterpillar's research people wanted to study how such a laser could be employed to cut and weld the tough steel alloys they use in building bulldozers and farm equipment. The laser we sold them produced "only" 10 kilowatts, in a continuous beam of invisible infrared energy, much like Wells's sword of heat. At that power level, it cut through three-quarter-inch spring steel at rates of fifty to a hundred inches *per minute*.

Although that laser consisted of a large roomful of heavy, bulky equipment, it still packed the power of Buck Rogers' faithful disintegrator sidearm. A space shuttle could lift that equipment in its capacious cargo bay and place it into a low orbit around the earth. The shuttle can carry thirty tons per mission; certainly two flights could have orbited that laser and its power generator. A 10-kilowatt laser is much too puny for ABM purposes, but it would make an effective weapon against satellites, which are light and fragile structures.

But I'm getting ahead of myself. Once the Defense Department realized that lasers of virtually any power output could be built, at least in theory, a two-pronged laser program was created. One part was devoted to developing high-power lasers of various kinds and testing them in the field. It was not enough simply to build such lasers and make them work; the military needed to know what those beams of energy would do to such targets as airplanes, missiles, re-entry vehicles, tanks, ships, and so on. The second part of the effort consisted of paper studies of how and where such weapons might best be used.

One fact became clear, I might even say transparent, right away: lasers would work better, and over much longer ranges, in the vacuum of space. On earth, the air we breathe presents resistance to the laser's beam of energy. The gas molecules in the atmosphere absorb some of the laser's energy, and the greater the distance between laser and target, the more energy absorbed. What is worse, as the air absorbs the laser beam's energy, it is heated and begins to form bubbles and waves, just as the heated air above a radiator shimmers and dances. This shimmering distorts the laser beam: instead of a straight line, the beam bends and warps with each motion of the air, and it becomes very difficult for it to hit a target. Not impossible, but very difficult. Today, nearly fifteen years after those earliest studies, lasers are being tested on both sides of the Iron Curtain for use by ground troops and naval vessels in anti-aircraft defense and other applications.

But in space, where there is no air to get in the way, a laser beam can be directed over very long distances with astronomical accuracy. And it is in space, where laser weapons can be turned against ballistic missiles, that lasers may find their most important military mission. Striking with the speed of light from orbiting satellites, laser weapons may be able to destroy ballistic missiles within minutes of their being launched, while their rocket engines are still burning and they are very vulnerable.

Alternatively, it may be possible to keep the lasers on the ground and have their beams reflected by mirrors that have been placed in space. This scheme would allow the lasers to be as large as necessary, but there would be no need to lift them into orbit. They could also remain within the territorial confines of their nation, where they would presumably be safer than they would be in orbit. Multifaceted mirrors, which the scientists call "segmented optics," might be able to adjust the energy beam coming out of the laser so as to compensate for the distortions caused by the atmosphere. Built on mountaintops, where the air is thin and clear, their powerful beams constantly adjusted by computer-controlled segmented mirrors, such lasers could direct enormously powerful energy beams to relatively cheap, easily replaced reflecting mirrors in orbit that are aimed by remote control to focus the lasers' energy onto the enemy's missiles.

Lasers are not the only weapons that might be applied to the ABM mission. Physicists have learned to accelerate streams of atomic particles, such as protons or electrons. For years such particle-beam accelerators have been used for pure research to probe into the nucleus of the atom and help us learn more about the fundamental construction of matter. But plowshares can be sharpened into swords, and in the 1970s rumors began to circulate that the Russians were developing a particle-beam weapon for use in space.

We know, from the scanty news about their space program released by the Soviets, that Russian cosmonauts have tested electron-beam devices aboard their Salyut space stations. So have we. An electron-beam device was among the scientific equipment aboard the first Spacelab mission in 1983. It was a Japanese scientific experiment, and it fired a stream of electrons into the earth's ionosphere, allowing scientists to study the behavior of electrified gases of the ionosphere under relatively controlled conditions. No military significance.

But Major General George J. Keegan, U.S. Air Force, became very concerned about Soviet development of particle-beam accelerators. As James Canan put it in his book, *War in Space:*

> High-energy weapons, their perils and their promises, began to penetrate the congressional consciousness in the late seventies, and it was . . . [General Keegan] — "crazy George" to his critics, "brilliant George" to his admirers — who started it all. Keegan undoubtedly harbored one of the very highest IQs ever to grace the military establishment, which is, stereotypes to the contrary, saying quite a lot. Keegan was in Air Force intelligence for many years, and in charge of it, starting in 1972, for five. He built it up to such size and scope that it became a *bête noire* for the CIA, which Keegan accused of intruding clumsily in military intelligence matters. . . . Keegan kept telling the civilian satraps in the Pentagon and in the CIA things they did not want to hear or believe, all under the heading that the Russians were coming. They devoted considerable time and energy to trying to prove him wrong and bad-mouthing him behind his back.

Keegan became convinced that the Soviet Union was developing a high-power particle-beam weapon for use in space, either as an antisatellite or as an antimissile weapon, or perhaps for both purposes. His main evidence consisted of reconnaissance satellite photographs of a sprawling research center at Semipalatinsk in Kazakhstan, and electronic communications concerning the work going on there, intercepted by "ferret" satellites. Other intelligence analysts, looking at the same data, agreed that the work being done at Semipalatinsk was aimed at producing a high-power particle-beam accelerator — but they concluded that the device was designed for research in particle physics, not space weaponry.

Keegan ultimately resigned from the Air Force and took his charges to the public. The media, with a few rare exceptions, paid him little attention. "A sizable segment of the U.S. scientific community jumped all over Keegan," wrote Canan, "accusing him of trafficking in paranoia." To this day, no resolution to the argument has been found. Are the Soviets developing a particle-beam weapon to fire at American satellites and/or ballistic missiles? Keegan still believes they are, and research has shown that a stream of subatomic particles could cause more damage to a hydrogen bomb–carrying re-entry vehicle than a powerful laser beam would. Professor Gregory Benford of the University of California at Irvine physics department (he is also an award-winning science fiction author) sums up the differences between a laser and a particle beam this way:

> Lasers . . . need to strike the same spot on the warhead for a full second in order to disable it. This means that the laser weapons will have to track the booster, not just snap a shot at it. Also, the first effect of heating a target [as a laser beam would do] is to make a cloud of hot, ionized gas, called a plasma, which quickly begins to reflect the laser beam, destroying most of its punch. A beam of particles, however, can bring to bear much more energy than the most powerful blowtorch, punching fist-sized holes in a microsecond, and can pierce through a plasma. These facts argue for using particle beams to penetrate hardened boosters [and warheads] and lasers for frying satellites in orbit, where lasers can focus on their targets for several minutes at a time.

As we shall see shortly, there is a third kind of weapon that can be employed against satellites and missiles: solid bits of metal, rather like glamorized, space-age versions of ordinary shotgun pellets.

While the technology of energy-beam weapons matured through the 1960s and 1970s, the fundamental concept of utilizing the "high ground" of orbital space for active military operations met with a vast amount of skepticism for several decades. And here we hit a fundamental difference in outlook among the experts, a decisive split not merely of opinions, but of basic attitudes as to how best to defend the United States against potential enemies while at the same time preserving world peace — or, at least, deterring nuclear war.

Since the end of World War II, when the United States had the atomic bomb and no one else did, the bedrock of American defense policy has been that we can destroy any nation that dares to attack us. Although nuclear weapons have proven to be useless in every other military and political situation, they have in fact prevented the two superpowers from going to war with each other. The consequences of nuclear war are so horrifying, and so well known, that the superpowers have refrained from using even tactical nuclear weapons on battlefields like Vietnam and Afghanistan — so far.

The Eisenhower administration's key phrase was "massive retaliation," meaning that an attack on the United States or any of its major allies would provoke a full nuclear strike against the aggressor. By the time Kennedy came to the White House, it was clear that the Soviets were working feverishly to build up an ICBM force that could reach the United States and obliterate most of our cities, causing casualties of a hundred million or more. Soviet missiles could also knock out American missile and air bases, leaving the United States powerless to retaliate.

Out of these realizations came the concept of "assured destruction." In *The Evolution of Nuclear Strategy*, Lawrence Freedman writes:

> No single public figure has influenced the way we think about nuclear weapons quite as much as Robert S. McNamara, the US Secretary of Defense from 1961 to 1968. . . . While he was in office many new concepts were introduced, of which the most important were *assured destruction, damage limitation,* and *flexible response,* which remain central to this day to strategic debate.

Assured destruction, simply defined, means that the United States decided to create strategic nuclear forces that are so powerful that the nation could absorb a first-strike attack by the Soviet Union and still have the wherewithal to destroy totally the USSR. In the parlance of prize-fighting, America could take the other guy's best shot and still knock him out — although, to carry that analogy one step farther, at the end of the bout both fighters would be on the canvas, not merely knocked out, but quite dead.

Even as early as 1946 it was clear that nuclear bombs were more political tools than military weapons. They were so powerful that they could not be used, at least not the way other weapons were. In Washington (but not, apparently, in Moscow) the defense establishment began recruiting a new type of strategist, the university-bred intellectual who began to plot the future course of nuclear weapons development and strategic planning. The think tanks began to proliferate, partly because scientific research and planning had been so obviously a major factor in winning World War II, partly because the military officers knew that they were armpits-deep in strange dark waters and they needed the help of experts. Men like Bernard Brodie, Herbert York, the late Herman Kahn, and many others began to study, write, and argue about nuclear policy — often in secrecy imposed by Washington. Occasionally their thoughts would come to the public's attention, as when Kahn published *Thinking About the Unthinkable* in 1962.

Fred Kaplan's book, *The Wizards of Armageddon,* is a fascinating examination of this "nuclear elite," from its beginnings in the Truman administration to the present day. Like other studies of nuclear strategy and the men who spend their lives thinking about the unthinkable, Kaplan's shows that every time the strategists have tried to maneuver away from a dependence on massive nuclear destruction, they have inevitably come straight back to that policy, like a ball on a rubber band bounding back to the place where it started.

An example. Early in the history of NATO, the strategists wanted to build up the alliance's conventional ground and air forces so that a land attack by the Warsaw Pact nations would not inevitably escalate into a nuclear confrontation between the United States and the Soviet Union. That failed, largely because the West Europeans felt that unless America's homeland was "under the gun" of Soviet missiles, the United States would be tempted to abandon her European allies in a crisis and return to isolationism.

A second example. Originally, the targets of our strategic bombers and missiles were Russian cities. Under this *countervalue* policy, we were prepared to kill a hundred million or more civilians if the Soviets attacked us. Analysts were uneasy with this genocidal policy and tried to work out a *counterforce* system, whereby our bombs would be aimed at Soviet weaponry, like military bases. But such a policy would mean that either we must absorb a first strike that would obliterate the nation or we must ourselves strike the Soviets' military machine before they strike us. In either case, American cities would be targeted and vulnerable, and the ultimate threat against a Soviet "city-busting" attack would be an American "city-busting" attack — like it or not.

And a third. The concept of an ABM defense against nuclear attack was formally rejected in the ABM Treaty of 1972, partly because the technical experts felt that a viable ABM system could not be built, but also because the strategists realized that an *effective* ABM system would destabilize the balance of terror between the superpowers. As York and Wiesner pointed out in their *Scientific American* article in 1964, it was felt that if one side possessed an "airtight" ABM system, it would be overpoweringly tempted to launch a first strike at the other superpower, knowing that its enemy's counterstrike would be intercepted.

In *The Wizards of Armageddon* Kaplan concludes:

> In 1946, in the beginning, Bernard Brodie wrote, "Everything about the atomic bomb is overshadowed by the twin facts that it exists and that its destructive power is fantastically great." The

story of nuclear strategy, from that moment on, has been the story of intellectuals — trying to outmaneuver the force of those axioms, trying to make the atomic bomb and later the hydrogen bomb manageable, controllable, to make it conform to human proportions. The method of mathematical calculation, derived mainly from the theory of economics that they had all studied, gave the strategists of the new age a handle on the colossally destructive power of the weapon they found in their midst. But over the years, the method became a catechism, the first principles carved into the mystical stone of dogma. The precise calculations and the cool, comfortable vocabulary were coming all too commonly to be grasped not merely as tools of desperation but as genuine reflections of the nature of war.

It was a compelling illusion. Even many of those who recognized its pretense and inadequacy willingly fell under its spell. They continued to play the game because there was no other. They performed their calculations and spoke in their strange and esoteric tongues because to do otherwise would be to recognize, all too clearly and constantly, the ghastliness of their contemplations. They contrived their options because without them the bomb would appear too starkly as the thing that they had tried to prevent it from being but that it ultimately would become if it ever were used — a device of sheer mayhem, a weapon whose cataclysmic powers no one really had the faintest idea of how to control. The nuclear strategists had come to impose order — but in the end, chaos still prevailed.

The nuclear strategists themselves would deny that charge, but they would be forced to agree that American — and Soviet — nuclear policy rests on the same foundation of sheer terror that it has always rested on. "Push me far enough," say the superpowers to each other, "and I will destroy you utterly, even though it means that you will destroy me, too."

The early policy of Assured Destruction grudgingly gave way to the concept of *Mutual* Assured Destruction, as the Russians worked tirelessly to equal, and then surpass, the American nuclear arsenal. As O'Keefe puts it in *Nuclear Hostages*, "When we arm, they arm. When we stop, they keep arming." The acronym for Mutual Assured Destruction is, of course, MAD, an inviting target for those who wish to criticize or lampoon nuclear policy.

Four decades after the explosion of the first atomic bomb, the world lives in the grip of nuclear terror, and the best policy that Washington and Moscow can implement appears to be MAD. The key to this policy is that the United States and the Soviet Union each offers its entire civilian population as hostages to the other side. Each side says to the other, "I dare not attack you, because I know your counterstrike will wipe out my people." Attempts to defend cities with ABM systems, or to make people safer through civil defense preparations, are regarded as destabilizing and counterproductive to the mutual balance of terror. Under the MAD policy, you and I are hostages; our lives depend on the good will and the good sense of the leaders in the White House and the Kremlin.

There are alternatives to MAD. One, of course, is nuclear disarmament. The United States could unilaterally dismantle its nuclear arsenal, or dismantle the bombs after working out an agreement with the Soviet Union for mutual disarmament. For decades the Russians have pressed for treaties that would aim at *total* disarmament, conventional as well as nuclear. The West has never trusted such Soviet overtures, however, because the Russians always refuse to allow international inspection or controls of any sort to be attached to their disarmament proposals.

Suppose the United States disarmed and Soviet Russia did not? At best, America would become a minor power, all of Europe would tie itself economically and politically to the USSR, and Russia would probably dominate Asia as well, although China would pose something of a threat to total Soviet hegemony in the Far East. Middle Eastern oil would flow to the United States only if the Soviets allowed, and at prices set by Moscow. Japan's vast consumer-oriented industrial power would be turned to selling automobiles and electronic gadgetry to Russians. American grain would still be sold to the Russians — at prices dictated by Moscow. That is most likely the best of all possible worlds if the United States disarmed unilaterally. Americans would be living under the constant threat of Soviet military power, which would be unopposed by American strength. Like Vichy France, our nation would no longer be independent or free.

If unilateral disarmament is unwise, what about a negotiated mutual disarmament that is lived up to by both sides? Again, Soviet intransigence over inspection and verification blocks the way. Certainly the people who long for a nuclear freeze have identified a good first step toward disarmament. But a nuclear freeze must be mutual and verifiable if it is to be workable. It took years of patient negotiations with the Russians to produce the limited success of the SALT I agreements. It took another decade to get SALT II signed by Carter and Brezhnev, and then the U.S. Senate gave every indication that it would refuse to ratify the treaty. When the Soviets moved into Afghanistan it was something of a godsend to Jimmy Carter: he found a face-saving reason to withdraw SALT II from the Senate's consideration. The START talks, the human rights talks, the European missile discussions — all of these negotiations drag on for years and are subject to the fits and starts and latest twists of policy or internal politics in both Washington and Moscow.

In fairness, though, Moscow and Washington can move quickly when the incentives are right. The Limited Nuclear Test Ban Treaty was essentially drafted in two weeks, in 1963, because both superpowers wanted to take a clear step away from nuclear confrontation after the Cuban missile crisis. In general, however, treaty negotiations move painfully slowly, because each nation perceives that it is dealing with the life and death of its people, and neither side trusts the other.

It may be possible to negotiate effective treaties with the Russians. It may be possible to move toward disarmament. But that movement will be slow, glacially slow, and it will happen only when each side is willing to take a certain amount of risk — something that is very hard to justify when you are dealing with the survival of your nation, your way of life, and the very lives of your people.

8
The Speech of 23 March 1983

THERE IS another alternative to MAD. It has been called *Assured Survival.* Perhaps the first man to use that term was Dr. Jerry E. Pournelle, a former military officer who has degrees in physics, psychology, and political science, and is a science fiction writer of wide popularity and an outspoken critic of MAD.

It may seem odd that so many writers of science fiction have been involved in strategic analysis. Someone whose only acquaintance with science fiction comes from watching motion pictures or television may suspect that the ideas being espoused are crackbrained, and that the media are perfectly justified in using "Star Wars" to describe such futuristic schemes as orbital ABM defense. But the truth is that, since the days of H. G. Wells, serious writers of science fiction have spent their careers examining the possibilities of the future. Long before "futurism" became a profession among scientists and a new area of advice to government and industry, writers like Wells, Olaf Stapledon, and Arthur C. Clarke were using science fiction to sketch out possible scenarios for the development of technology and the human consequences of such developments.

The idea of Assured Survival rests on the possibility of defending the nation against nuclear attack. This means that a way must be found to defend against ballistic missiles. One of the earliest proposals along these lines was made during the Eisenhower years. It was dubbed BAMBI (for ballistic missile boost intercept, not the innocent little fawn in Felix Salten's book and the popular 1942 Disney movie). BAMBI called for hundreds of satellites, in low earth orbit, armed with projectile guns that would fire solid shot at approaching ballistic missiles.

Pournelle, who believed that one of the advantages of BAMBI was that it would be cheaper than MAD, found that the top brass in the Pentagon and their civilian advisers thought that putting up several hundred satellites was a ridiculous idea. More than that, the idea ran into grave political opposition from the Strategic Air Command.

"You could see even in those days," Pournelle says, "that those were your alternatives: either you were going to have to go to defense or you were going to have to start building more and more [ICBMs]. . . . The SAC generals wouldn't even discuss defense; they saw defense as a way to take their bombardment capabilities away from them. And they still do!"

In a debate on the issue of space militarization held at the 41st World Science Fiction Convention, in Baltimore, on 2 September 1983 (yes, serious issues are discussed — vigorously — by the science fiction people), Dr. Robert Bowman, former director of advanced space programs for the Air Force, reinforced Pournelle's point. He opposed the idea of space-based defenses against missile attack, saying that "every dollar spent on defense can be neutralized by five cents of offense," and that he would favor a treaty banning all weapons in space. He attacked the Star

Wars ABM concept, saying that it would not work — but later suggested that orbital defenses might be useful in support of an American first strike against the Soviet Union, where the ABM satellites would defend the United States against a Russian retaliatory blow. This, he said, is precisely what the Soviets fear and why they see an American proposal for orbital ABM satellites as a direct threat to them.

In the 1960s, Pournelle was the general editor of a top secret Air Force study of strategic doctrine and missile technology, "Project 75," and then became principal investigator for a study titled "Stability and Strategic Doctrine," conducted for the Air Force and presented to the Air Council. Later, while he was a professor at Pepperdine College in the Los Angeles area, he and Stefan T. Possony coauthored a study later published under the title *The Strategy of Technology*, which has been used as a text in the Air Force Academy and the Air War College. The book again raised the concept of Assured Survival based on orbital ABM defense. Pournelle became a consultant to the National Security Council during the early years of the Nixon administration, and *The Strategy of Technology* was brought to the council's attention by one of its leading members, Richard Allen, who was later to become President Reagan's first National Security Adviser.

Shortly after Reagan became President and Allen moved into a basement office in the White House's west wing, Pournelle helped to organize the Citizens' Advisory Council on National Space Policy, an ad hoc group of scientists, engineers, writers, activists, and others who were eager to promote a stronger American space program. The Advisory Council had no official capacity whatever; it was "sponsored" by the L5 Society, a national grass-roots activist organization whose goal is the creation of permanent colonies in space. In this case, as Pournelle put it, sponsorship meant merely that "the L5 Society paid the Advisory Council's bills." The council was in no other way connected to the society.

By that time, the early 1980s, there was a plethora of space activist groups, most of them quite small, each of them pushing for a slightly different goal, even though all of them were agreed that the American space program was not moving far enough or fast enough to suit them. Trying to get them to work together was about as difficult as trying to get the Democratic Party to agree on a single candidate for President.

But Pournelle was able to organize the Citizens' Advisory Council, with its enthusiastic but argumentative and sometimes divided membership, to the point where it produced a significant report, urging that space technology be used to change basic American strategic doctrine from MAD to Assured Survival. That report found its way to Richard Allen in the White House. By the time Allen was replaced by William Clark, the basic tenets of Assured Survival were already making an impact on the Oval Office.

Meanwhile, similar ideas were being offered to the White House by a retired general. "Because I'm in favor of a space-based defense," Dan Graham says cheerfully, "everybody assumes that I was in the Air Force." Lieutenant General Daniel O. Graham, U.S. Army (Ret.), was an infantry officer for most of his career. He was chief of intelligence for General William Westmoreland through much of the Vietnam conflict, became a deputy director of the Central Intelligence Agency, and then was named director of the Defense Intelligence Agency.

In 1981 Graham, under the aegis of the Heritage Foundation, published a study titled *High Frontier,* which laid out a plan for an orbital ABM system that electrified the space enthusiasts.

"The origins of this effort lie back in the days when I

was a military adviser to then-candidate Ronald Reagan," Graham wrote in *High Frontier*. "Early in the [1980] campaign I was among those insisting that the only viable approach for a new administration, when coping with the growing military imbalances [vis-à-vis the Soviet Union], was to implement a basic change in the U.S. grand strategy — to make a 'technological end-run on the Soviets.'"

Graham saw two fundamental problems with existing American strategy, both stemming from the fact that the strategy, based on the MAD concept, forces the United States and the Soviet Union into a never-ending escalating arms race. The first problem is that this arms race is horribly expensive and will become only more expensive as time passes. The second problem is that, since the arms race is devoted to offensive nuclear strike weapons, it makes the world constantly more dangerous.

"Our defense policy does not defend us," Graham asserted. It threatens mass destruction as a deterrent to attack. Graham saw MAD as a policy of offering hostages, not as a defense. He made a distinction between MAD and what he called "true deterrence."

According to Graham, true deterrence "recognizes [that] you can persuade an attacker not to attack by showing him that his attack won't work. If his attack won't work, it's obvious he can't destroy you and you don't have a MAD situation."

By early 1981, Graham became convinced that the one area where the United States could make a "technological end-run on the Soviets" was in space. In the spring issue of *Strategic Review* he published an article, "Toward a New U.S. Strategy: Bold Strokes Rather than Increments," that presented the basic idea of a space-based ABM defense. He called the concept High Frontier, which caused a good deal of confusion and some resentment among space activists, because Princeton professor Gerard K. O'Neill's idea of building colonies in space, the huge permanent habitats that are the goal of the L5 Society, was originally published in a book titled *The High Frontier.*

Confusion and resentment aside, by autumn 1981 High Frontier had become a project of the Heritage Foundation, a conservative think tank, and Graham was able to bring together a distinguished group of scientists, military officers, and political advisers to flesh out his basic concept. High Frontier became a nonprofit corporation based in Washington, with the backing of such influential conservatives as Phyllis Schlafly, several representatives and senators, and prominent industrialists like Justin Dart, Joseph Coors, Jack Hume, and Karl Bendetson, who were close enough to President Reagan to be considered part of his "kitchen cabinet."

The major differences between the High Frontier proposal and the space-based ABM systems we have already discussed in this book is this: in an effort to produce a system that can be placed in orbit within five years, and at as low a cost as possible, Graham and his cohorts have tried to develop a system that uses "off-the-shelf" technology wherever possible. Thus, instead of depending on laser or particle-beam weapons, High Frontier would rely on small, high-velocity missiles that are fired at the attacker's ICBMs and destroy them by impact. In essence, the system calls for a very sophisticated version of bullets, rather than energy beams or nuclear explosives, to destroy the ICBMs.

According to Graham's study, "The total costs of the High Frontier system over the next five or six years, in constant dollars, might be roughly $24 billion. Through 1990, the total costs in constant dollars would probably be around $40 billion — a figure that compares favorably with what would have been the total cost of the MX . . . in its original configuration."

High Frontier envisions a "layered defense," consisting

of satellites that can fire at an attacker's missiles while they are boosting from the ground (or sea), backup satellites that can go after the missiles that survive the first layer of the defense, and finally ground-based "point defenses" at the target areas to knock out any missile warheads that get through the gauntlet in space. The orbiting satellites, collectively, are called the Global Ballistic Missile Defense; the High Frontier study calls for 432 such missile-carrying "trucks" in orbit to assure global around-the-clock coverage of any and all possible attacks. Each truck carries forty to forty-five small missiles that are pointed at their targets by sensors aboard the truck and then launched. Each layer in the three-layered defense should be capable of destroying 90 percent of the missiles it faces, which means that only 0.1 percent of the attacker's total force would reach its target.

To critics who point out that even 0.1 percent of a massive nuclear attack could destroy several cities and kill millions of people, Graham replies coolly that "you never have a perfect defense, not against the bullet, not against the tank, and not against nuclear weapons." A 100 percent effective defense is not necessary, Graham asserts. All that is necessary, he maintains, is to convince the Soviets that a nuclear attack on the United States would not achieve its objectives. That would forestall the attack. Deterrence would be served, because the Soviet strategists could not be certain that a nuclear attack on the United States would destroy our retaliatory capability or wipe out all our cities.

The most serious criticisms of High Frontier, of course, have concerned Graham's figures on costs and the time required to put the system into operation. It is probably a law of nature that proposals for radically new technological systems tend to underestimate the difficulties, especially the costs, of producing such systems. When you are faced with a skeptical, perhaps even a hostile, audience, you tend to put the best possible gloss on the product you are trying to sell.

But Graham asserts that his strongest opposition has been based not on his cost estimates or time schedules, but on High Frontier's being seen by his former colleagues in the Pentagon as a threat to their ongoing programs. If High Frontier were accepted by the White House and the MX program scrapped, for example, a large number of Air Force officers, civilian employees, and private contractors would lose their jobs. No doubt some Navy personnel see High Frontier as a threat to future programs for ballistic missile submarines, such as the Trident, and to shipbuilding programs.

Once Reagan came into the White House, his top advisers intensified their search for new initiatives in defense. Graham's High Frontier concept found an attentive audience in the Oval Office. As we have already seen, many of Reagan's close friends, industrialists and leading Republican Party figures like Senator Malcolm Wallop of Wyoming — became supporters of the High Frontier concept.

Meanwhile, the physicist Edward Teller, known as the father of the hydrogen bomb, was pursuing his own ideas of space defenses, based on x-ray lasers that would be energized by the explosion of small nuclear bombs. "Defense is the best deterrence," Teller believes, "if it works."

The Hungarian-born Teller is a man who has aroused strong passions among his colleagues. Ardently anti-Communist (perhaps anti-*Russian* would be a more accurate description), Teller was a central figure in the events that led to Robert Oppenheimer's dismissal from government service. Oppenheimer opposed development of the hydrogen bomb; Teller gloried in it. In 1952 Teller helped to create the Lawrence Livermore National Laboratory, under the aegis of the University of California; it soon became a leading center for research in nuclear power and weaponry.

While all the other schemes for space-based ABM defenses depend on non-nuclear weaponry, Teller came up with the idea of developing "third-generation" nuclear weapons, in which the energy created by the explosion of a small nuclear bomb would be used to generate powerful x-ray laser beams. Such beams of penetrating x rays could be much more effective in destroying missiles and warheads than laser beams in the lower wavelength regions of ultraviolet, visible, or infrared radiation. In a series of underground nuclear tests in Nevada, code-named Excalibur, Livermore scientists have apparently been successful in producing x-ray emission from a laserlike apparatus "driven" by a small nuclear explosion.

Long an advocate of active defense against ballistic missiles, and an opponent of the 1972 ABM Treaty, Teller served as a scientific adviser to Ronald Reagan in the 1976 and 1980 political campaigns. In September 1982 he made the case for his version of space-based ABM defenses to the President in a briefing at the White House. Admiral James D. Watkins, after further briefings from Livermore scientists, drafted a report on the idea and presented it at the President's regular meeting with the Joint Chiefs of Staff in February 1983.

At that meeting the President and the Joint Chiefs discussed the need for developing the new MX ballistic missile, which was being heavily opposed in the Congress, and the problems of assuring that the United States' ground-based missile force could survive a Russian "silo-busting" first strike.

Ronald Reagan came to the White House committed to strengthening the U.S. defense posture, which inevitably meant large increases in funding for the Pentagon. One of the key programs included in this buildup of American forces was the MX — the new, heavy, more accurate, ten-warhead ICBM that Reagan's advisers saw as the answer to "the window of vulnerability," when very accurate Soviet missiles would be able to destroy many of our Minuteman missiles in their silos.

The MX has had, to say the least, a checkered history. The original idea was to build a powerful new ICBM that could be moved around the countryside, from one launching silo to another, so that the Russians would not know where the missiles were from one day to the next, and thus could not aim their own first-strike missiles at our retaliatory ICBMs.

The Air Force had been promoting this idea since the late 1960s, realizing that improvements in Soviet missile accuracy made fixed missile silos increasingly vulnerable. By the late 1980s this window of vulnerability could mean that extremely accurate Soviet SS-18 and SS-19 missiles, with their very powerful warheads, might be able to wipe out most of our Minuteman missiles in their silos. Our manned bomber force would still be based on the fleet of aging B-52s; even with cruise missiles that could be released a thousand miles from their targets, the Air Force worries that the B-52s are very vulnerable to Soviet air defenses.

That would leave only one leg to the American strategic triad: the U.S. Navy's fleet of ballistic missile submarines, with their 568 Poseidon and Trident missiles, which carry a total of 5152 nuclear warheads. These nuclear subs, cruising for months at a time beneath the ocean's surface, are a mobile missile force that is extremely difficult for the enemy to find. If there is a window of vulnerability for American nuclear submarines, it will not open until well into the 1990s or later. But submarine-launched missiles are not yet as accurate as missiles launched from fixed

The Speech of 23 March 1983

sites on land; the Poseidon and Trident missiles could be used only for a countervalue city-busting attack, not a counterforce silo-busting strike.

To the Air Force, this was an intolerable situation; they could not abide leaving the entire nuclear strike mission to the Navy. Defense analysts agreed that the concept of a three-legged strike force — ground-based missiles, manned bombers, and submarine-launched missiles — was sounder than any one leg by itself. Thus was born the idea of a mobile land-based missile, the MX. It would be larger and more accurate than the Minuteman; it would carry ten MIRV warheads of 300 kilotons each — fifteen times the explosive power of the Hiroshima bomb. To ensure its survivability in the face of a Soviet first-strike silo-busting attack, MX was to be transported around hundreds of miles of "racetracks," from one silo to another, so that Russian surveillance satellites would not be able to tell which silos contained missiles and which were empty.

The same logic of upgrading American strategic nuclear forces in the face of growing Soviet capabilities also dictated that a replacement for the B-52 was long overdue; hence the B-1 bomber program and the cruise missile, which can be launched from the planes while they are still a thousand miles from their targets.

In the closing days of the Carter administration, it became clear that the people of Utah and Nevada, where the MX racetracks were to be laid out, wanted no part of the system. Even the conservative Republicans among them had no desire to invite a saturation attack of Russian nuclear bombs to their states.

Various schemes were examined for housing the missiles, including putting them in ships or submarines (anathema to certain Air Force people), flying them around the country in huge cargo planes, and basing them in a "dense pack" mode — the exact opposite of the original dispersed, mobile-basing scheme. Each of these schemes had serious drawbacks, so serious that none of them could be adopted. But the Reagan administration wanted to go ahead with developing the MX and building the first group of missiles.

There was another aspect to the MX debate, as well. Originally conceived as a movable missile that could foil a Soviet silo-busting attack, the MX almost inevitably came to be seen as a weapon that had the accuracy and firepower to attack Russian missiles in *their* silos in a counterforce strike. More than 80 percent of the Soviets' total nuclear strike capability rests in their silo-based ICBMs, unlike the United States' more evenly balanced strategic triad. If the Russians perceive the MX as a first-strike silo buster, they would undoubtedly try to find an answer to this new threat, thus escalating the arms race even more. Critics of the MX, including leaders of the Nuclear Freeze Movement, saw the new missile as an unnecessary and certainly unwanted additional turn in the spiraling arms race.

In the spring of 1983, the Scowcroft Commission, a panel of distinguished experts appointed by President Reagan to study the problems facing America's strategic forces, turned in a report that damned the MX with faint praise. The commission was chaired by retired Air Force general Brent Scowcroft, former National Security Adviser to President Ford, and included among its members Richard Helms, former head of the CIA, and Alexander Haig, former chief of NATO and Secretary of State. While calling for deployment of the first hundred MX missiles in existing Minuteman silos, the Scowcroft Report urged that the Air Force move toward development of a smaller, cheaper, single-warhead missile that could be more easily made mobile than the massive MX. Immediately dubbed "Midgetman" by the media, this new small missile would offer survivability without threatening another round of escalation in the arms race. The *New York Times* approved editorially of the Scowcroft Commission's conclusions, saying that "a prudent course for America's strategic weapons is finally coming into view." But *Science* magazine, the weekly journal of the American Association for the Advancement of Science, noted that "there is a surprising appetite for the cold war in most of the commission members."

At their meeting in February 1983 with the President, the Joint Chiefs discussed missile defense systems for protecting all American strategic missiles in their silos, the old Minuteman or the new MX. Reagan saw BMD in a larger context, as a means of protecting not only the missiles, but the cities and the people of the United States and its allies. More than that, a space-based defense might help to sell MX to the Congress.

In a way, then, it was the problem of the MX — as much as the ideas espoused by Teller, Graham, and the like — that moved President Reagan to include the Star Wars concept in his defense policy speech of 23 March 1983.

The President still wanted to get the MX construction started, whether or not the land-based nuclear forces moved on to Midgetman later in the 1980s. So in his speech he repeated the need to push for the development of the MX. He showed charts that displayed how the Soviets were outpacing the United States in missiles, warheads, and the other paraphernalia of nuclear destruction. And then he surprised everyone by suggesting what can only be described as a basic change in strategic policy — the long-urged shift from MAD to Assured Survival.

> My advisers, including in particular the Joint Chiefs of Staff [said President Reagan], have underscored the necessity to break out of a future that relies solely on offensive retaliation for our security....
>
> Would it not be better to save lives than to avenge them? Are we not capable of demonstrating our peaceful intentions by apply-

> ing our abilities and our ingenuity to achieving a truly lasting stability? I think we are — indeed, we must!
>
> After careful consultation with my advisers . . . I believe there is a way. Let me share with you a vision of the future which offers hope.
>
> It is that we embark on a program to counter the awesome Soviet missile threat with measures that are defensive. Let us turn to the very strengths in technology that spawned our great industrial base and have given us the quality of life we enjoy today.
>
> Tonight, consistent with our obligations under the ABM Treaty and recognizing the need for close consultation with our allies, I am directing a comprehensive and intensive effort to define a long-term research and development program to begin to achieve our ultimate goal of eliminating the threat posed by strategic nuclear missiles. . . .
>
> I call upon the scientific community in our country . . . to give us the means of rendering these nuclear weapons impotent and obsolete. . . . Is it not worth every investment necessary to free the world from the threat of nuclear war? We know it is!

Most of the nation, most of the world, indeed most of Reagan's own advisers, were startled by this announcement. Many White House advisers had not been informed that the President was going to go public with the missile defense idea. Others were decidedly opposed to the concept. The confusion was evident during the next few days, as the media tried to learn just what the President had in mind.

Apparently the President had personally drafted the Star Wars part of his speech, and then invited a few of his closest advisers to review it. Among them were George A. Keyworth II, the President's Science Adviser and head of the Office of Science and Technology Policy (OSTP). "This was not a speech that came up [from lower echelons]," Keyworth said. "It was a top-down speech . . . a speech that came from the President's heart. We were told what to do by the President."

According to *Science* magazine, after the February 1983 meeting with the Joint Chiefs, no additional scientific analyses were done in preparation for the speech, although Keyworth consulted with Solomon Buchsbaum and William O. Baker, both of Bell Laboratories and members of the White House Science Council. On the evening of 23 March they and eleven other scientists were invited to the White House for dinner. Among the others were Nobel laureate Hans Bethe; the former head of the Defense Department's research and engineering division, John Foster; Frank Press, president of the National Academy of Sciences and formerly President Carter's Science Adviser; the scientist-industrialist Simon Ramo, the R in TRW Corporation (and rumored to have been Reagan's first choice for Science Adviser); Charles Townes, spiritual father of the laser; and Teller. Also present at the dinner were Keyworth, Under Secretary of Defense Fred Iklé (Defense Secretary Caspar Weinberger was out of the country), and national security aide Robert McFarlane — soon to be elevated to National Security Adviser.

The President's dinner guests were surprised by his revelation. Most of them were reserved in their responses; the general impression was that they were more concerned about the political impact of a space-based defense than about the scientific or technical difficulties of creating orbital ABM systems.

The reaction to the Star Wars speech was immediate and predictable. Veteran strategic analysts, the "Wizards of Armageddon" that Kaplan and others have written about, were jarred to their bones. For decades their calculations had told them that Mutual Assured Destruction depended on each side's refraining from building missile defenses. Deterrence depended on holding the other side's civilian population as hostages. Defense of any kind was a threat to MAD, a threat to deterrence and the balance of terror. A defense that was based on hundreds of satellites firing "death beams" — it was scoffed at, called impossible, immoral, and at the very least a violation of the ABM Treaty.

Meg Greenfield wrote in the *Washington Post*, "Maybe nuclear stability would be threatened by the president's initiative. But *certainly* nuclear orthodoxy has been threatened by his enunciation of it."

Senator Edward Kennedy castigated the President's speech as "misleading Red scare tactics and reckless 'Star Wars' schemes." Soviet Premier Yuri Andropov warned that if the United States began to place ABM satellites in orbit, it would "open the floodgates to a runaway race of all types of strategic arms, both offensive and defensive."

Jerome Wiesner, President Kennedy's Science Adviser and former president of MIT, said that "it's really a declaration of a new arms race." Richard Garwin of IBM said flatly, "It won't work." Ramo, who was generally enthusiastic about the President's announcement, cautioned, "We don't know how to do this yet, and there will be pitfalls and problems along the way." He also pointed out that technology developed for defense might also become useful for offense. Bethe expressed fears that "we are going right into space wars. We will be in serious trouble if these systems work."

One of the supporters of space-based defenses, anticipating the hesitancy of the scientists, reportedly chided the President after his speech, "You shouldn't have asked the scientists to do this; if you want a missile defense right away, ask the engineers. They know how to get things done."

Although the world buzzed with commentary on the Star Wars theme, the main emphasis of Reagan's speech was support for his $244.5 billion defense budget in general, and procurement of the first hundred MX missiles in particular. The prospect of a defense against nuclear attack was clearly held out as a future ideal, "a new hope for our children in the twenty-first century." Space-based ABM defense was the sugar coating; the MX and the defense budget were the pill.

Whatever his motives, and whatever time the program takes for development, the fact is that Reagan's speech marked the first step away from MAD and toward the concept of Assured Survival. A Defensive Technologies Studies team, headed by James C. Fletcher, former NASA administrator, was appointed to examine the technological aspects of space-based ABM defense and make recommendations to the Department of Defense and the President. Another group tackled the strategic implications of space-based defenses.

Fletcher's group quickly found itself in a welter of conflicting technical and political pressures. General Graham and his High Frontier cohorts urged a program that could be implemented immediately. Graham is convinced that within five years the United States could have missile-armed satellites in orbit that would begin to offer some protection against a Soviet first strike. At the very least, they would be able to destroy Russian satellites, thereby providing a counter to the Soviet ASAT system.

Other scientists are investigating the use of "rail guns," electrically powered catapults that would sling small dart-like projectiles toward oncoming missiles at incredible velocities; the darts would shatter missiles or warheads on impact. The redoubtable Teller urges that the best way to destroy ballistic missiles is with his x-ray-emitting directed nuclear weapon. But it is the chemical laser that is best understood and could be tested in space the soonest. Chemical lasers draw their energy from the chemical reactions of two or more gases, such as fluorine and deuterium. However, chemical lasers emit energy in the infrared region of the spectrum, not the best wavelength for damaging or destroying targets. Researchers are now working on *excimer* lasers, which use noble gases, like argon and krypton, together with such halogens as fluorine, and "lase" in the far more energetic ultraviolet wavelengths. And there is also the possibility that particle-beam weapons will turn out to be better missile-killers than any laser.

Then there is the question of whether the ABM weapons should be placed in orbit, where they will be vulnerable to enemy attack, or kept on the ground until needed, at which time they would be "popped up" to assume their defensive stations in space. Or perhaps only reflecting mirrors will be orbited and huge, enormously powerful ground-based lasers will bounce their beams off them and onto attacking missiles.

George Keyworth, the President's Science Adviser, has become convinced that "the major assets of a strategic defense system need not be space-based." He envisions massive laser installations on the ground, using computer-controlled segmented mirrors to direct their powerful beams through the atmosphere without distortion and then to reflect the beams off orbiting mirrors. The mirrors in space can be as large as football fields, if necessary, and as lightweight and inexpensive as a Mylar sheet coated with a thin spray of metallic reflecting material.

As the enthusiasm for starting a space-based defense effort gained momentum through the summer and autumn of 1983, other events combined to push Reagan's defense budget and the embattled MX program through the thicket of congressional opposition. While the White House and Pentagon were insisting that the MX was necessary for the nation's defense, and would even help the American position in disarmament talks with the Soviets, opponents of the new missile system in the Congress were arguing that no suitable basing mode had been formulated; and without a basing scheme that would make the missile relatively invulnerable to a Soviet first strike, MX was little more than a costly and needless escalation of the arms race. The Star Wars concept was also being heavily attacked by various critics as destabilizing, too expensive, or physically impossible.

Then, on 31 August 1983, a Soviet SU-15 fighter plane shot down Korean Air Lines flight 007, killing all 269 passengers and crew aboard. The 747, on a routine flight from Anchorage to Seoul, had strayed into Soviet airspace over the Kamchatka Peninsula, where a major Russian base for ballistic missile submarines exists near the city of Petropavlovsk. The Soviet air defense command apparently confused the civilian airliner with a U.S. Air Force electronic "snooper" plane, an RC-135, which is also a four-engine jet similar in general outline to a 747, although considerably smaller. Apparently the Soviet fighters dispatched to intercept the "intruder" did not inspect it visually and could not hail it by radio. They fired tracer bullets past the plane in an effort to attract the pilot's attention, Moscow later claimed, and when the plane failed to respond, one of the fighters fired a pair of heat-seeking missiles. The airliner crashed into the cold Sea of Japan.

Worldwide indignation was loud and insistent, for a few days. At the United Nations, American Ambassador Jeane J. Kirkpatrick denounced "this calculated attack on a civilian airliner." President Reagan called it a "massacre" and a "crime against humanity." Most of the furor quickly died down, but during the storm of anger against the Soviets, the Congress passed Reagan's defense budget — including the MX.

Reagan was given what he wanted. By November 1983 a joint House-Senate conference committee approved a compromise $249.8 billion defense appropriations bill. It was $11.1 billion lower than the White House's amended budget request, but it included $2.1 billion for the first twenty-one MX missiles, $1.7 billion for the Trident missile program, and $5.6 billion to procure the first ten B-1B strategic bombers. Reagan finally had won approval to build the first MX missiles, even if they had to be based in old Minuteman silos. His arms control negotiators could face their Soviet counterparts from the position of strength that Reagan had insisted was necessary.

And he had also won approval for the concept of a space-based ABM defense. Or at least he had the funding to start examining the possibilities in greater detail.

It would take political steps of unprecedented magnitude to create a truly global missile defense system, a defense system that protects *every* nation on earth against attack by *any* nation. Neither the United States nor any other nation is ready to take such steps today. But in another decade, when space-based weaponry may make missile defense a reality, such steps could be the difference between Assured Survival and World War III.

10

Smart Weapons

I SAT BEHIND the pilot and watched over his shoulder as he put the sleek, speedy plane through its paces. High above the Mohave Desert, the experimental jet soared and looped as the pilot worked the controls. I could see the flat horizon tilting and twirling as the plane performed.

"Flutter!" somebody shouted as the plane's stubby wings began to shake violently. Before either of us could respond, the left wing shattered into a dozen fragments and the plane snapped over onto its back, then plunged straight down toward the desert floor.

The pilot took his hands off the controls, rubbed them on his knees for a moment, then stood up and walked away from the cockpit. The television screen in front of him went blank. I remained seated for a moment longer; then I too got up and followed the pilot out of the cockpit area.

The experimental plane, named HiMAT (highly maneuverable aircraft technology) crashed onto the desert floor. But its pilot and his visitor (me) remained in the ground-floor office from which the plane was controlled.

In that office is a fully instrumented cockpit for the HiMAT plane. The pilot sits in a desk chair rather than an airplane seat, and he can wear his street clothes. No need for an oxygen mask or a parachute. He sees, thanks to television cameras mounted in the plane's real cockpit, exactly what he would see if he were inside the actual cockpit. He handles the plane with a joystick, rudder pedals, throttle, and other controls, just as he would in flight. The instruments in his cockpit faithfully report what the plane is doing, instant by instant. But he is sitting in an office building, many miles away from the plane he is "flying."

For several years NASA has been flight-testing HiMAT aircraft at the Dryden Research Center, which is located at Edwards Air Force Base on Rogers Dry Lake. The place is drenched with "the right stuff." Here Chuck Yeager broke the sound barrier in the rocket-driven X-1 on 14 October 1947. The X-15 rocketplane flew here, from 1959 to 1968, probing the uppermost limits of altitude (67 miles) and speed (4520 miles per hour) that an airplane could achieve, piloted by men such as Scott Crossfield, Joe Engle,[*] and the man who would eventually be the first to set foot on the moon, Neil Armstrong. And here the space shuttle made its first landings.

HiMAT is a gorgeous little airplane, built about half the size it would have to be if it actually carried a human pilot. Jet-powered, its swept-back wings located toward the tail, and rakish-looking canard airfoils forward of the main wings, it is a sweet-looking ship. NASA and the Air Force jointly use HiMAT to test new aerodynamic shapes, lightweight composite structural materials, and new electronic control systems. Because modern electronics allow the pilot to remain safely on the ground, the plane can be made

---

[*] Engle is still an active astronaut and has flown the shuttle.

smaller and cheaper than it would otherwise be. And it can be put through flight maneuvers that would be too dangerous for even a test pilot to undertake.

In the jargon of the engineers, HiMAT is *tele-operated.* That is, the plane is controlled remotely. Electronic links between the pilot sitting in that office building and the plane flying high above the desert allow the pilot to handle the plane just as if he were aboard it. Computer scientists like MIT's Marvin Minsky believe that the principles of tele-operation will someday allow people to control all sorts of machines over distances of hundreds, thousands, even millions of miles. Imagine sitting in a sunny office in Santa Barbara and controlling a mining machine digging coal out of a pit in West Virginia. Or operating a deep-sea dredging machine from the comfort of your own home. Or, from the safety of a control station miles away, handling a robot that is inspecting the intensely radioactive core of a nuclear reactor. Or, from a control center in Houston, directing a spring-wheeled exploratory rover as it rolls across the red sands of Mars.

Tele-operation will have many uses in the future. For the military, it offers the possibility of removing human troops from the battlefield. For years now the Pentagon has been working toward the electronic battlefield, a complex system of sensors and computers that will allow commanders to see exactly what is going on at the front, from half a world away. Combining the electronic battlefield and tele-operated weaponry leads to the opportunity for a roboticized army.

I do not — repeat, *not* — mean that we will see battalions of human-shaped robots clanking across a battlefield like a stainless steel version of Pickett's charge. There is no reason to build robots in human shape, especially when the only reason that humans are put into battle is that there has been no other way to carry weapons to the enemy effectively — until now.

Think about it. The purpose of an infantryman is to observe the situation that his commanders want to know about, and, when they decide that he should do so, to use his weapons to kill, chase, incapacitate, or cow the enemy into surrender. All the panoply of the modern army — attack planes and helicopters, tanks, artillery, armored personnel carriers, trucks, and everything else — are for the support of the infantryman. It is he, with his five senses and the integrating power of his intelligence, who decides which way the battle goes.

Until now, there has been no way to replace that scared young man carrying a rifle or automatic weapon toward the enemy's frightened young man. But his basic functions of observing, reporting, and delivering firepower do not necessarily require that he be physically present at the battlefield. Not if sensors and weapons can be developed that do not need a vulnerable human being to carry or employ them.

You can start a vicious argument among computer scientists by asking them when, if ever, computers will become as intelligent as humans. But we do not need computers of human-quality intelligence to replace men in battle. Today's computers are almost good enough; tomorrow's computers will undoubtedly fill the requirements quite completely. We do not need "intelligent" computers. In the terminology of the technologists, we already have "smart" computers; the latest machines are even described as "clever," and those now on the drawing boards are being called "brilliant." Such computers, in robots or other of machines, can replace human soldiers, sailors, and men in battle. If war is too important to be left to the generals, then it is far too dangerous to be left to any human being — draftee or volunteer — to do the fighting.

To observe, to report information back to the commanders, and to employ weapons: those functions of the fighting man can be taken over by machines. Some of them will be tele-operated, directed by human beings from safe distances. Others will be smart or even brilliant weapons, automated suicide machines, electromechanical kamikazes that can find their designated targets and destroy them by flinging themselves upon the foe. Some machines like that already exist. They are called drones, remotely piloted vehicles (RPVs), guided missiles, and smart bombs.

The first attempts to make drone aircraft came during World War I, when primitive radio equipment was mated to a few biplanes, with results that hardly were noticed, except by a few forward-thinking engineers. During the zany years between the two World Wars, when fliers were doing all sorts of barnstorming stunts to earn a living and keep on flying, more work was done on controlling pilotless airplanes remotely, from the ground. On 24 September 1929, Lieutenant James H. Doolittle, U.S. Army Air Corps, took off from Mitchel Field on Long Island, New York, in the rear cockpit of a two-seater biplane. His cockpit was completely enclosed by a hood that cut off all view of the outside. Guided only by a radio link to the ground and the instruments glowing feebly inside his darkened cockpit, Doolittle flew the little plane to a thousand feet, and after about fifteen minutes turned it around and made the first "blind" landing in aviation history. Doolittle went on to become a champion air racer, the leader of the first bombing raid on Tokyo in World War II, commander of the powerful Eighth Air Force, and, after the war, an aerospace industry executive. But his pioneering "flight in the dark" was his most important contribution to the development of aviation and *avionics* — which is the engineers' word for the electronics systems of aircraft and missiles.

Arthur C. Clarke, long before he became a world-renowned author, worked on the first radar systems for ground control of aircraft landings. In fact, his first published piece of writing was about those shaky early experiments in GCA (ground-controlled approach); it was titled "You're on the Glide Path . . . I Think."

From blind flying, relying on instruments or ground controllers rather than eyesight, to controlling a plane remotely from the ground, is mainly a matter of avionics. Today planes can be controlled completely, from takeoff to landing, by tele-operators on the ground. Commercial airliners do not need pilots aboard for routine flying. The crew in the cockpit is there because the customers, the safety-conscious Federal Aviation Agency, and the pilots and their unions do not trust fully automated flying. But the day will undoubtedly come when that confidently smiling man in the cockpit will be little more than a figurehead, placed there to reassure the passengers that a dashing yet fatherly human being is ready to take charge if any emergency arises. The real pilot will be a computer with a human monitor sitting in an office somewhere on the ground.

The Germans developed a primitive kind of remote-controlled weapon during World War II. To attack Allied shipping from the air, they armed their long-range bombers with small unmanned airplanes that were released by the bomber and controlled by a radio operator aboard the "mother plane" to dive into a targeted ship. The most widely used of these radio-controlled flying bombs was the Hs-293, which was slightly more than eleven feet long and weighed about twenty-three hundred pounds. It was powered by a small rocket engine and directed by the bombardier aboard its mother plane. Very effective against Allied shipping, the Hs-293 could be used only in clear weather, where the human tele-operator could see both the drone and its intended target. These flying bombs were among the earliest electronically controlled "standoff weapons" — weapons that allow the human crew to remain out of range of the enemy's defenses while they launch their attack. (The very earliest standoff weapon of written record was, I suppose, David's rock-throwing sling, which felled Goliath long before that bruiser could get his powerful hands on the young shepherd.)

In 1944, a radio-controlled German missile called Fritz-X, guided from the bomber that released it, sank the Italian battleship *Roma* after Italy had surrendered to the Allies. Most of the other ships sunk by German drones during World War II were freighters and tankers, in convoys bound for Britain or the Soviet Union.

World War II also saw the development of drone weapons, such as the famous — or notorious — German V-1, a small, pilotless jet bomber that flew blindly until it ran out of fuel, then carried its one ton of explosives down to a fatal crash dive. It came to be called the "buzz bomb," because of the distinctive putt-putt sound of its pulsed jet engine. Londoners quickly learned that as long as you could hear the buzz bomb, you were all right. It was when its engine quit and it began its final plunge to earth that you were in trouble.

"George Orwell, author of *1984,*" writes David Ritchie in his book, *Space War,* "became familiar with the ominous silence that preceded a buzz bomb's landing. . . . He noted that the brief delay gave one just enough time to hope the bomb hit someone else — and to reflect on the 'bottomless selfishness' of the human race."

More than 8000 V-1s were fired at London alone, and although they killed and maimed thousands of civilians, they made little impact on the war. The buzz bombs were inaccurate, unreliable, and slow enough for Allied fighter planes such as the Spitfire and Mustang to shoot them down.

For the V-2 rockets, however, there was no defense at all, except to destroy or capture their launching pads. The V-2 was probably the first weapon system to use a true ballistic guidance system. Gyroscopes and associated electrical equipment sensed the rocket's motion through space and activated control vanes in the V-2's exhaust plume to alter its course if and when it strayed from its planned trajectory. But because the rocket engines burned for only a minute or so, the course corrections were quite crude, and the V-2's accuracy was very low. Missile accuracy is described in terms of circular error probability (CEP), a hypothetical circle, drawn with the target at its center, into which 50 percent of the missiles launched at that target will theoretically fall. The V-2 had a CEP of a hundred miles or so; the Nazis launched it in the general direction of London and hoped for the best. Today's ICBMs have CEPs of a few hundred *yards*.

Modern ballistic missiles, of course, are the very epitome of push-button warfare. But they are decidedly "all or nothing" weapons. Once launched, they cannot be recalled. And once launched, they invite the enemy's immediate counterstrike. In a very real sense, ballistic missiles that carry hydrogen bomb warheads are not weapons of war; they are political symbols of national determination. If they are ever used, it will not be to wage war, but to annihilate a nation, a continent, perhaps the whole world.

While ballistic missiles are the heart of the superpowers' strategic nuclear forces, modern technology has developed smaller, cheaper weapons that can be used more flexibly for either tactical or strategic purposes — weapons that remove the human being from the fighting.

The V-1 and radio-controlled flying bombs of World War II have evolved into the standoff missiles and cruise missiles of today. I remember, back in the 1950s, when the Air Force was trying to develop the Snark, a pilotless jet plane of intercontinental range. Most of the Snarks launched from Cape Canaveral somehow never got much farther than the beach before crashing, and we newcomers to the cape were warned not to go swimming in the ocean, because the waters were "Snark-infested." One Snark actually did fly its full multithousand-mile range, but because a technician had not replaced a faulty element in the guidance system before takeoff, instead of cruising out to Ascension Island in the mid-Atlantic, the plane crashed into the Mato Grosso jungle in Brazil, stirring an international incident; the Brazilians did not appreciate having Yankee bombers dropping into their territory.

Missiles are much more sophisticated today. The cruise missile, which can be launched from the ground, from the deck of a ship, from a submerged submarine, or from a plane, is a very smart robot airplane indeed. Pilotless, jet-powered, its guidance system allows it to hug the ground as it zips along almost at the speed of sound, following the terrain as it dips into hollows and rises to clear hills, matching the ground whizzing by below it with its internal map of the path it must follow to its target.

The United States has developed basically two cruise missiles. The Boeing Corporation builds the air-launched cruise missile (ALCM) for the Air Force, which identifies it as AGM-86, the AGM standing for air-to-ground missile. The ALCM is designed to be carried by the Strategic Air Command's aging B-52s, and will also be carried by the B-1 bombers when they enter active service. It is smaller than all but the tiniest of sport planes, slightly more than twenty feet long, with a wingspan of twelve feet and a total weight of 2825 pounds. It has one turbofan engine, which can get the ALCM up to about 500 miles per hour. Its range is 1550 miles.

General Dynamics Corporation's Tomahawk, AGM-109, has been chosen by both the Army and the Navy and is therefore known as both the sea-launched and ground-launched cruise missile (SLCM and GLCM). In its GLCM mode, it is the cruise missile that NATO is now deploying in Western Europe. In size and performance, it is similar to the ALCM.

Cruise missiles can carry conventional or nuclear warheads. They can be launched from the land, the sea, under water, or from planes. Their fifteen-hundred-mile range means that the launching platform — and crew! — can stand well away from enemy defenses. Their speed, their ability to fly very low and hug every contour of the terrain, and their low radar profile means that they are hard to find, to track, and to intercept. The cruise missile is a very flexible kind of weapon, adaptable to a wide variety of military missions. That is what makes it such a threat to the opposition; that is why it is such a sticking point in arms control negotiations.

For the Air Force, the cruise missile is looked on as a standoff weapon that allows the crews of old B-52s and brand-new B-1s to fire hydrogen bombs at the USSR without coming within range of Soviet air defenses. This is particularly important for the B-52s, which originally entered the Strategic Air Command in 1954. The "latest" B-52s in SAC were built in 1962. Russian supersonic jet fighters with heat-seeking air-to-air missiles and radar-guided antiaircraft missiles and artillery have had more than twenty years to catch up to the B-52s' capabilities. The cruise missile gives the bomber crews a chance to do their jobs and get back alive. The Air Force plans to buy more than three thousand by 1987.

Future developments in cruise missiles will see them becoming smaller, faster, smarter, and harder to detect. "Stealth" technology, which uses new materials and designs that reduce the radar and infrared "signature" of an aircraft, can make a small missile virtually impossible to find. It has been estimated, for example, that stealth technology can reduce the radar reflection of the 150-foot length of the B-1 to the amount of reflection a one-square-foot plate would give. Cruise missiles, and smaller aircraft, can become invisible to radar and heat-seeking infrared sensors.

The Nuclear Freeze Movement in Europe started when the United States announced that it planned to place 464 GLCMs and 108 Pershing II medium-range missiles in Europe, over a period of five years, to upgrade NATO's tactical nuclear forces. This was a response to the Soviet Union's decision to put more than two hundred SS-20 medium-range missiles (which carry three warheads apiece) into Eastern Europe and the western part of the USSR. The SS-20s, which have sufficient range to hit London from launching pads in the Ural Mountains, are, say the Russians, merely replacements for aging, obsolete missiles that have been aimed at NATO targets for decades. But the Soviets fear the Pershing IIs, because they can reach targets inside the Soviet Union in ten minutes' flight time. And they fear the cruise missiles, which present a new kind of threat because of their flexibility and relative cheapness as compared with bigger rocket-boosted missiles.

On the tactical battlefield missiles have become crucially important. Francis Scott Key wrote about "the rockets' red glare" during the British bombardment of Fort McHenry in the War of 1812. Genghis Khan's Mongol hordes used rockets back in the thirteenth century. In World War II the U.S. Army employed the bazooka anti-tank weapon, and the Russians and Germans also developed various forms of rocket artillery.

But it was not until the 1970s that it became quite clear

that tactical missiles were assuming primary importance on the battlefield and at sea. In the Arab-Israeli War of 1973, the Yom Kippur War, which led to the Arab oil embargo against Europe and the United States, infantrymen using shoulder-fired antitank rockets took a fearsome toll of enemy armor. It was not uncommon to see tanks running away from infantry. Less than nine years later, in the Falkland Islands War of 1982, Argentine planes damaged and sank several British ships, including the destroyer *Sheffield*, with air-to-surface missiles.

The battlefield missile has come of age. Lone attack planes, even lone infantrymen, now pose a very credible threat to huge ships and massive tanks. The pride of the armies and navies — their menacing, ground-shaking tanks and their billion-dollar aircraft carriers — are now threatened by relatively inexpensive missiles that strike faster than the speed of sound. The U.S. Navy, for example, is so concerned about the threat to its aircraft carriers from small missiles carrying nuclear warheads that it is spending billions of dollars to develop "close-in" defenses to shoot down those missiles if (and when) they get past the combat air patrols and picket ships arrayed around the carriers. Future shipboard defenses may even include high-power lasers, whose energy beams strike with the speed of light.

The missiles are getting smarter every year. During the final years of America's futile involvement in Vietnam, the Air Force began using low-power laser beams to spot a target for a smart bomb — a bomb with movable fins that allowed it a measure of control after it had been dropped from an airplane, and with an optical sensing system that could find the brilliant gleam of laser light and direct the bomb's control system toward that signal. The laser essentially painted a sparkling bull's eye on the target, and the bomb homed in on it.

Infantrymen now have laser-designating missile systems

to use against tanks, bunkers, and other "hard" targets. The main problem with them is that the soldier must hold the laser beam on target until the missile hits. This may take only a few seconds, but even that short time can be fatal if the enemy sees the soldier. So the Defense Department is working on "fire and forget" missile systems that will allow the soldier to take quick aim at a target, fire his missile, and then duck back under cover. The missile will be smart enough to recognize the target and get to it. On the other hand, why have a soldier risk himself at all, if a pilotless drone airplane can spot the target and "paint" it with a laser beam?

Drone aircraft — either remotely piloted vehicles or unmanned vehicle systems — have entered the battlefield. These pilotless planes are used for missions of reconnaissance, targeting, and electronic warfare. Why send a human pilot on a reconnaissance mission over enemy territory when an RPV can do the job? With today's advanced electronics and sensors, the human pilot is merely a ferryman guiding a flying platform of observation equipment. Instead of placing him in danger, the commander can use a pilotless plane. And once the human is removed, the airplane itself can be made smaller and cheaper, which means it will be harder for the enemy to find and destroy, and less expensive to replace if it is destroyed.

Reconnaissance RPVs are designed to be used over and over, of course. Some of them are so small that they look like the model airplanes children fly in the park on a summer afternoon.

The Aquila mini-RPV is manufactured by Lockheed Missiles and Space Company for the U.S. Army. Designated YMQM-105, the Aquila is only six feet, six inches long, with a twelve-foot, six-inch delta-shaped wing that turns down at the tips. It is a propeller-driven airplane, powered by a twenty-six horsepower two-stroke engine, like the kind of motor used in lawnmowers. Weighing only 220 pounds, Aquila can stay aloft for up to three hours. It can carry television cameras and other sensors, and is used to give tactical commanders a picture (literally) of the battlefield. Aquila can find targets, point laser beams at them for smart weapons, and then give an assessment of the damage done.

Israel is a nation of small population surrounded by more numerous foes. The Israelis have developed RPVs to a high art, and even used RPVs as radar-jammers against Syrian surface-to-air missiles (SAMs) in the Bekáa Valley when they invaded Lebanon in 1982.

The original military use of drone aircraft was serving as targets for human gunners. In the earliest days of World War II, fliers learned air-to-air gunnery by shooting at a target sleeve towed behind a plane. Many a towplane pilot felt himself in such danger from the wild shooting of inexperienced airmen that combat against the enemy was almost an anticlimax. Today, exceedingly clever drones can be programmed or tele-operated to fly the way a Soviet plane would be expected to fly in combat. Instead of merely popping away at a target sleeve, today's fighter pilot flies against as good a simulation of his potential enemy as modern technology can produce. Drones like the Beechcraft MQM-107 are reusable, inexpensive, and offer realistic target practice. They can carry radar and infrared countermeasures similar to the jammers and flares that an actual enemy plane would use. The rocket-propelled AQM-37 drone can sprint to speeds of Mach 3.5 and altitudes of more than ninety thousand feet.

These smart, clever, brilliant drones, missiles, and tele-operated equipment are being used for reconnaissance, over-the-horizon targeting, and for the crucially important, interlinked functions of battlefield command, control, and communications (which the military calls C³). Army leaders today speak of the "electronic battlefield" and technology concepts for the "Air/Land Battle 2000."

Unmanned vehicle systems, their proponents point out, are cheaper than the targets they help to destroy. They save dollars as well as the lives of the soldiers and airmen they replace. The UVSs offer other advantages, as well. NATO commanders plan to use cruise missiles to replace manned jet attack planes that are now held for nuclear strike missions against targets deep inside the Warsaw Pact nations. Relieved of their "deep interdiction" missions, these planes and crews can now be used as non-nuclear fighters or attack bombers at the battlefront. The cruise missiles, then, will act as a "force multiplier"; in effect they will increase the number of planes NATO has available to fight at the front.

The concept of force multipliers is foremost in the thinking of NATO and U.S. military leaders. The U.S. Defense Department is pushing for better standoff and tele-operated weapons as a counter to the Soviets' numerical advantages in troops, tanks, artillery, and planes. Dr. Robert S. Cooper, director of the Defense Advanced Research Projects Agency (DARPA), put it this way:

> We must get to a position where either through [force multiplying] technology or through the production of large numbers of military equipment and a buildup of our fighting forces, we are able to oppose the Soviets. We must be able to convince them that a conventional attack against us would be met by force they could not overwhelm. [We should have] something other than the threat of nuclear weapons to deter conventional conflict with the Soviets.
>
> Currently, about the only way that we have of assuring this is to build up the [conventional forces] that we have, and we are doing that in a substantial way. But we cannot hope to match the Soviets. They have . . . such a large fraction of their effort [in] . . . producing military equipment. . . . We are not going to match them aircraft for aircraft, tank for tank, or ship for ship.

Smart Weapons

Cooper believes that America must either devote a huge percentage of its gross national product to defense, or develop new technologies to offset the Soviet preponderance of numbers.

> Fortunately [he continued], I think we have technologies which are nascent now, but which are coming of age. . . . I believe that smart weaponry . . . standoff weaponry that has advanced sensors and intelligent computing machines built right into the weapons, hold the potential of being able to neutralize very large massed forces . . . with a very small opposing force. Because the probability of "kill" for the weapon can be driven up to almost one hundred percent and the ability of the opposing forces to stop those weapons . . . can be almost completely negated.

Smart weapons can stabilize what Cooper and other defense analysts see as an increasingly unstable situation, particularly in Europe, where growing Warsaw Pact forces face shrinking NATO forces:

> The massing of forces makes it possible . . . in a localized area, to gain sufficient superiority to invite . . . localized conventional war. So, if you can provide technical capabilities that make it unlikely that a military force would succeed, even with the massing of [such] forces . . . then there would be deterrents even of localized conflict . . . and that could ultimately result in stopping localized conflicts.

Certainly within ten years it will be possible to develop brilliant missiles directed remotely by tele-operators or guided by microchip computers no larger than a fingernail. They will be cheap enough to build and deploy by the thousands, even the tens of thousands. They will be deadly enough to make the battlefield too dangerous for human beings. Together with the advent of ABM satellites, they offer the key to warfare suppression.

An International Peacekeeping Force armed with such smart, cheap missiles could make any battlefield in the world untenable for human beings, at very little risk to its own human operators and decision-makers. The IPF's people will be far from the battlefield. Most of them will be in deep underground command posts, safe from anything short of a direct nuclear hit. A few will be manning planes or ships hundreds of miles from the battlefield, from which they launch their drones and missiles.

Cannot the aggressor also use brilliant machines, and launch a completely automated attack? Perhaps so. But attack is always more difficult than defense; the attacker has more choices to make, more problems to solve. Yet even if the aggressor goes to a totally automated force, what of it? The IPF's task is really quite a simple one: destroy the weapons being massed for the attack, or — once battle is joined — destroy the weapons being used on both sides. Like a policeman called in to stop a barroom brawl, the IPF is not concerned with the reasons that the fight started; the IPF's job is to *stop the fighting*. If it can be stopped before it even begins, fine. If not, then both sides are going to be hit by the policeman. It will be up to the courts to decide the right and wrong of the matter; the policeman's job is merely to preserve the peace.

PART III

Dialogues

And, of course, there was the atomic bombing of Hiroshima and Nagasaki, which in our view was not the last salvo of World War II, but rather the first one heralding the Cold War.
— Georgi Arbatov

The willingness to reassess and move beyond an earlier formulation is what makes one's actions a model. This testifies to the ability to outgrow parochialism, to move beyond the starting point, to give up the comfort of having a fixed position.
— Hyman G. Rickover

11

Scenario Two: The European Face-Off

TOWNS IN CENTRAL EUROPE are approximately 2 kilotons apart. That is, an explosion equal to 2000 tons of TNT cannot take place anywhere in Central Europe without damaging at least one town. Since nuclear warheads, even the smaller tactical or so-called battlefield weapons, are considerably more powerful than 2 kilotons, a nuclear war cannot be fought in Central Europe without enormous devastation to the civilian population.

Still, the standard operational credo of NATO's forces calls for using tactical nuclear weapons against a Soviet attack. This is based on the assumption that Russia and her Warsaw Pact allies of Eastern Europe hold an overwhelming advantage in conventional forces: tanks, planes, artillery, and troops. Many Western military experts dispute this, claiming that NATO estimates of Warsaw Pact strength have always been exaggerated, and that the two antagonists have roughly equal power, all things considered. Even so, they concede that the preponderance of sheer numbers is solidly on the side of the Soviets.

To counter those numbers, NATO doctrine calls for

immediate use of battlefield nuclear weapons. The Russians know this, and are prepared to use their own nuclear weapons as a form of superartillery to obliterate the NATO forces opposing them.

Two world wars have been fought on European soil. World War II saw the equivalent of several million tons of TNT unleashed across the continent. Cities from Rotterdam to Stalingrad were smashed flat. More than forty million European men, women, and children were killed. Yet World War III could erupt there at almost any moment. Like the Second World War, the Third could well begin in Poland.

Tensions across Poland reached their snapping point in the spring. It had been a long, cold, hungry winter. The Polish economy, plagued by low productivity and a lack of foreign credit, was racked by the workers' sullen slowdowns that not even the sternest measures of the military government could contravene. Stores were virtually empty of goods. Even the long-flourishing black market offered little more than potatoes and bread. Solidarity, long outlawed by the generals who ran the government, still lived in the minds of the people. And the Roman Catholic Church still offered its maddening blend of patience and resistance to the generals.

It was inevitable that rioting would break out with the first warm weather of spring. Women waiting in food lines for hours grew shrill and angry when the shopkeepers revealed that there was no meat, no fresh vegetables, no milk, not even flour. In Gdansk, in Warsaw, in a dozen other cities, they marched through the dirty gray snow that still lined the streets, not back to their homes, but to the factories and offices where their husbands worked.

"There is no food!" they screamed from the streets. "Why work when there is no food?"

Scenario Two: The European Face-Off

The factories and offices emptied. The army was caught by surprise. The streets were suddenly filled with angry men and women, smashing windows, overturning cars, tearing down government banners, and toppling statues of their erstwhile leaders. By the time night fell, fires blazed in every major city across the land. From their headquarters in Warsaw, the generals escalated the battle against their own people, giving orders to use riot police, water cannon, tear gas, and — by midnight — regular troops with live ammunition against the rioting crowds.

By dawn it became apparent that the army had split apart. Reports streamed in to Warsaw that the troops in the streets were refusing to fire on the people. In several places the troops turned their guns against the riot police, instead.

The message from Moscow was brutally direct: the Red Army will restore order. As the Russian tanks moved from bases in the Soviet Union and East Germany, the capitals of Western Europe reacted with frenzied calls back and forth across the Atlantic. In Washington, the newly inaugurated American President issued a carefully worded warning, insisting that the Soviet Union "immediately stop its interference in the internal affairs of Poland," and threatening "appropriate actions by the nations of the free world" if the Red Army did not turn back.

Moscow replied that it was acting in response to a request for aid from the Polish government, and the generals in Warsaw duly made that request. But the fuse was lit and burning. It was only a matter of time before it reached the powder keg.

It happened in Germany, near the town of Bad Neustadt in the hilly, wooded area known as the Thüringer Wald. The mass rioting in Poland, after years of military repression, had sent shock waves throughout Eastern Europe. Hungary and Czechoslovakia seethed with memories of

Russian tanks rumbling through their capitals. Many tried to take the opportunity to flee to the West. Whole families bolted for the borders. Nowhere was the unrest more pronounced than in East Germany. Overnight a vast exodus began, Germans of every age and social status driving, hitchhiking, walking toward the border that divided East from West, seeking refuge with their fellow Germans.

It was at the border, not much more than a mile from Bad Neustadt, that a jittery East German soldier fired on a handful of civilians as they threaded their muddy way through the barbed wire and minefields that separated the two Germanies. From observation posts in the woods on their side of the border, West German soldiers fired machine guns in the air as a warning to the East German troops to leave the civilians alone. Within minutes a regular firefight was going on, with the civilians pinned down in the middle of it. Within hours, West German soldiers were on East German soil, actively helping their brethren across the border.

Moscow warned that this was a deliberate provocation and threatened immediate retaliation. West Berlin was sealed off by a cordon of Russian tanks, and the Soviet Premier announced that the presence of West German troops in East Germany was an act of war. He gave the Bonn government twenty-four hours to recall its troops *and to return all the East German citizens who had fled into the West.*

NATO began mobilizing. Troops were called up from the reserves in all the nations of Western Europe. The United States activated its war-status airlift, and huge cargo planes, laden with troops, supplies, and ammunition, began roaring off from airfields all across the eastern third of the United States. The U.S. Navy went on full wartime alert, sending battle units to predetermined "choke points" in the North Atlantic and the Sea of Japan.

Scenario Two: The European Face-Off

At NATO headquarters in Brussels, and deep within the Kremlin, confusion and doubt raced side by side. Although both military staffs had had plans of attack and defense laid out for some three decades, no one had envisioned a situation like this. The Red Army could not count on its Warsaw Pact units; indeed, the Russians had their hands full trying to pacify Poland and keeping the other Soviet bloc nations from erupting into rebellion. But the threat to West Germany was real and immediate.

"Can we fight a limited campaign in Germany?" the NATO generals asked one another.

Before they could come to an answer, the Russians struck with devastating force.

Twenty nuclear-armed missiles, the SS-20s that had precipitated the nuclear freeze crisis of the early 1980s, blasted up from their mobile launching platforms in western Russia. Ten minutes later, sixty key NATO bases disappeared into the rising mushroom clouds of radioactive gases that dotted Western Europe. Airfields, supply depots, troop bases, shipyards, and military headquarters in five nations were wiped out in the flash of a millisecond. Several million civilians who happened to make their homes near the military targets were killed by the blast and firestorms. Whole towns were obliterated. Major cities, among them Brussels, Hamburg, Essen, Amsterdam, and Milan, were heavily damaged. Millions of people suffered lethal doses of radiation and died within a few weeks.

Its "decapitation attack" hugely successful, Moscow immediately offered to negotiate a truce, pointing out that it would be impossible now for NATO to defend Western Europe against a conventional Russian ground assault. The Soviets had refrained from striking France or Britain — or the United States, a point that was not lost on the civilian leaders of the NATO nations. Only nations that had no nuclear retaliatory power of their own had been attacked.

But American and British troops had been killed, by the thousands. Although the French held their forces within their own borders, they too were shocked and furious at the Soviet attack. The pressures within the United States and Great Britain were not for negotiation with the Russians; far from it. Even with the threat of nuclear devastation hanging over them, the people overwhelmingly demanded that their governments "give the Reds a dose of their own medicine."

After a frenzied few hours of nearly hysterical meetings, the President took the unprecedented step of calling the Soviet Premier on the hot line and broadcasting the conversation live on national TV. But even as the President demanded that all Russian troops leave Poland *and* East Germany, Soviet killer satellites were launched by the dozens and swiftly snuffed out most of the American early-warning and reconnaissance satellites in low orbit around the earth.

"We do this as a precaution and a warning," the Soviet Premier told the President. "We know that you are preparing to attack the Soviet Union. We are prepared to defend ourselves."

"Tell him," the President said, white-faced and with trembling voice, to the Russian interpreter, "that our response will be measured and just. We have no desire to kill Russian civilians, but we cannot allow the Red Army to gain control of Western Europe."

The Premier responded that the fate of the human race was in the President's hands. Either he could accept the situation that now existed in Europe or he could destroy the world. "It is up to you" were his final words.

The American response was measured and just, as the President had promised. Fifty cruise missiles, launched from submarines in the North Sea and from long-range bombers, streaked across the skies of Europe to seek out

Scenario Two: The European Face-Off 201

their preset targets of Warsaw Pact military bases. But the tanks and guns and planes they were programmed to destroy were no longer at those places; they had been moved days earlier. Blinded by the loss of their reconnaissance satellites, the Americans fired their symbolic volley anyway. Some 5 megatons of nuclear fury erupted across Eastern Europe, killing millions of Poles, Czechs, East Germans, Hungarians, Bulgars, and Rumanians — but very few Russian soldiers.

Still, the Russians responded with a limited city-busting attack that obliterated Cologne, Antwerp, and Bologna. Even before the warheads had exploded on their targets, a dozen Trident missiles from American and British submarines popped out of the Atlantic Ocean to destroy military bases within the Soviet Union. The next exchange of missiles devastated Britain and France. The final exchange — the full ICBM forces of both sides — left the northern half of the globe under a pall of radioactive cloud for many weeks.

When the sun broke through the gray overcast at last, the only humans left alive in the northern hemisphere were a few scattered families living deep in mountain or forest retreats. They called themselves survivalists, but they had at best only a few years' worth of canned food. Radiation poisoning killed most of them long before their food ran out. Disease and starvation took care of the rest as a nuclear winter closed its dark fist over the remainder of the globe and slowly but remorselessly obliterated all human life.

If an International Peacekeeping Force existed, could it ward off World War III? Would it be able to oppose directly the armed might of one of the superpowers — or both of them? That is a little like asking how far is up. An IPF might be able to forestall a nuclear war between the superpowers, especially if it had the strength and the credibility to intervene successfully at a lower level of confrontation.

Suppose the IPF were a creature of the smaller nations of the world. Suppose NATO and the Warsaw Pact still existed, still maintained their military might, but the IPF was in the process of placing laser-armed satellites in orbit and organizing its army of remotely controlled robot weaponry.

When the rioting begins in Poland and the Red Army mobilizes, the IPF issues a warning to Moscow that any military intervention in Poland will bring on retaliation. The Kremlin responds that the Red Army is preparing to restore order in Poland at the request of the Polish government, and any hostile action by the IPF "will elicit the appropriate response."

But within the Kremlin's carpeted conference rooms, an argument rages: What is the appropriate response? The IPF is not a nation; it is a multinational organization. Its headquarters is in Geneva. Its military bases are scattered throughout a dozen small countries, most of them in the third world. The United States sells weaponry to the IPF but is not yet a formal member. Denmark and Norway quit NATO to join the international force; Sweden was a charter member of the IPF. Several other NATO nations are on the verge of joining. To nuke Geneva is unthinkable: the whole world would come down against the Soviet Union. Attacking IPF bases in Sweden, India, Malta, or elsewhere would be just as counterproductive.

After two days of debate, the answer comes to the surface. If the IPF attempts to intervene, knock out their satellites. Blind their orbiting observation posts and eliminate their antimissile satellites. The task is assigned to the Strategic Rocket Forces.

The Red Army begins to move into Poland, only to be met by clouds of cheap, swift, smart drone missiles and pilotless airplanes that swoop in on the rumbling tanks,

Scenario Two: The European Face-Off

destroying hundreds within the first hour of combat. Launched from small ships in the Baltic Sea and from long-range planes that were based in Malta and Algeria, the remote-controlled kamikazes force the Red Army spearheads to halt their advance into Poland.

The Strategic Rocket Forces have already launched their antisatellite satellites the day before, establishing them in orbits that could easily be matched to the orbits of the IPF's observation and ABM satellites. But before the ASATs can be maneuvered into their proper positions, the IPF's laser-armed satellites begin zapping them into harmless blobs of melted metal. Within an hour all the Soviet ASATs are destroyed, and the IPF lasers turn to the Russians' own surveillance satellites, destroying them also with quick strokes of invisible laser energy beams.

It is the Russians who are blinded. Two warnings issue from IPF headquarters in Geneva. One, to Moscow, that if the Soviet Union does not disperse its military forces away from the Polish borders, the next targets of the satellite lasers will be the USSR's communications satellites. Two, to Washington, that if the United States attempts to take this opportunity to threaten or attack the Soviet Union, the ABM satellites will deal with the American missiles just as they dealt with the Soviet satellites.

A Soviet submarine torpedoes one of the IPF's missile-launching ships in the Baltic, killing most of the international crew. Swedish antisubmarine boats and planes, operating under the IPF flag, find and sink the submarine half an hour later; all hands go down with the sub.

Both the United Nations Security Council and General Assembly meet in emergency sessions. The General Assembly quickly passes a resolution, by unanimous voice vote, condemning "armed intervention against a member state." In the Security Council the Soviet ambassador indicates his government's willingness to settle the Polish question by peaceful means. In Geneva, representatives of Hungary and Rumania quietly begin negotiations to join the International Peacekeeping Force — as do Italy, the Netherlands, and Greece.

The threat of Soviet missiles has not been eliminated from the calculations of European and American strategists, but it has been greatly diminished. In the Kremlin, several ranking hawks of the Politburo offer their resignations, accepting graceful retirement in return for their lives. The Soviet ambassador to the United States hints that the Soviet Union may be willing to consider joining the IPF and reducing its armed forces if the United States will do the same. Before his State Department has a chance to weigh the relative advantages and drawbacks, the American President uses the hot line to tell the Soviet Premier that the United States would welcome such a move.

If that scenario seems too fanciful, too naïvely optimistic, remember that the Limited Test Ban Treaty — the first treaty between the United States and Soviet Union that allowed the superpowers to take a step away from impending nuclear holocaust — came to pass only after the Cuban missile crisis of 1962 had brought the two nations teetering on the brink of Armageddon. Remember too that the SALT agreements and all subsequent arms-limitation and/or arms-reduction negotiations depend heavily on the presence of surveillance satellites in orbit that can count the number of missile silos and submarines each nation possesses.

The truth is, however, that no International Peacekeeping Force exists, except for the token units of troops thrown into the cauldron of the Middle East, where their mission is mostly symbolic and they do not have the numbers, the weapons, or the authority actually to enforce peace.

Scenario Two: The European Face-Off

The first steps toward an effective IPF have already been taken, though. Not by the politicians and diplomats, but by the scientists and engineers. Much of the hardware necessary for ABM satellites and drone weapons already exists. Surveillance satellites have been with us for twenty years. Smart missiles and high-power lasers are being tested in the field today.

What does not yet exist is the "software," the political mind-set that would allow an IPF to come into existence. In the next few chapters we will examine the reactions of decision-makers on both sides of the Iron Curtain to the realization that laser-armed ABM satellites can be placed in orbit. For it is this step, or, rather, the perceptions in Washington and Moscow of how this step can be realized, that will determine whether we live in peace or die in a world-engulfing nuclear holocaust.

12

No. 14 Borodin Avenue, Kutusow Prospekt

From his living room on the top floor of the highrise apartment block Viktor Ilyich Aleksandrov has not merely one, but two beautiful views of Moscow, as befits a ranking member of Gosplan, the state planning bureau. From one side of the spacious and handsomely furnished room you can see the Moscow River, broad and busy with barges and sightseeing boats. By turning your head and looking out the other set of windows, you can see Red Square and the imposing towers of the Kremlin itself.

Aleksandrov's five visitors are all gawking at the city, a panorama of twinkling lights in the darkening evening. Their host has not offered them drinks or food, but they don't seem to care. The view and the luxury of a four-room apartment for one man alone seems to have stunned them all.

"I asked you here," Aleksandrov tells the five men, "to discuss a very grave matter."

The visitors glance at each other, their faces showing more curiosity than uneasiness. Three of them are old friends and comrades of Aleksandrov, highly placed in their respective branches of the military and government. The other two are much younger, but well known as men on the rise. All of them are solid members of the Communist Party.

"An important matter?" asks Boris Grechko. "What is it, your latest love affair?"

Aleksandrov smiles tolerantly and the others allow themselves a quiet laugh. Viktor Ilyich is well known for his various amours. He is not the handsomest of men; his nose is larger than it should be and his chin is rather weak. But he can be quite charming when it is profitable, and he knows how to dress and dine in style. Beneath his suave exterior is a soul of steel: hard yet supple, cold and ruthlessly sharp. His smile is famous among the women, but his ice-blue eyes never let down their guard.

Grechko is his oldest friend, an overweight, overwrought section manager in the planning bureau who at one time was Aleksandrov's boss. Grechko eats too much, smokes too much, and talks too loudly. But he survives. Aleksandrov is not the first ambitious young man to climb over Grechko's big belly on the road of advancement. But Viktor Ilyich is the only one who is still in Moscow; the others all faltered, one way or another, and are either far out in the hinterlands or gone for good.

"No, my love life is doing quite nicely at the moment, thank you," Aleksandrov says. "What I have in mind is this latest American problem — their idea for putting up satellites to shoot down our ballistic missiles."

There is a long moment of silence. Grechko paces solemnly to the plush sofa set along the far wall of the living room and sits down heavily. Aleksandrov motions to the others to be seated.

"What I want from you," he says, "is a full exchange of your ideas on the subject. You can speak frankly here; nothing that you say will be written down."

"So that you can take credit for our good ideas," grumbles Grechko.

"Of course." Aleksandrov does not need to say that his apartment is not bugged or that the security police will not record their conversation. They all understand.

Grechko and two of the others are old enough to have lived through the Nazi invasion of the Great Patriotic War. Colonel General Vladimir Koshkin was a teen-aged partisan when the great Battle of Kursk took place, the largest clash of tanks in the history of warfare. He personally killed four Germans with a grenade he threw into their tent the night before the battle started. By then, both of his parents and his two sisters had been killed by the German invaders. Academician Dmitri Maslov spent the war behind the safety of the Ural Mountains, a young laboratory assistant working with the USSR's greatest physicists on rocket propellants. Most of his family died in the siege of Leningrad. Grechko himself, an infantry private for most of the war, came home to find that his wife had been killed in a bombing raid and his three sons had disappeared. Like millions of others, he never found his children.

The two younger men are Nikolai Birman, an economist whose daring ideas have caught the attention of the new First Secretary, and Major Mikhail Roznenko of the Strategic Rocket Forces.

"None of you will talk unless I bribe you with liquor, is that it?" Aleksandrov asks, smiling his bright smile. "I don't want this discussion to turn into a drunken songfest."

"It's not that, Viktor Ilyich," says Maslov, defensively. "It's just that the subject is — well, to be frank, it is very disturbing."

"Why else would I ask your opinions?"

Maslov is the oldest man in the room, nearly seventy, bald except for a fringe of dead-white hair. After the war he was assigned to work with Igor Tomm and Andrei Sakharov on the atomic bomb project.

"I don't think it can be done," says Colonel General Koshkin. "Putting death rays in satellites and shooting down all our missiles. Nonsense!"

Aleksandrov, still standing by the window with the lighted turrets of the Kremlin behind him, arches his brows.

"It is possible in theory," Academician Maslov says softly.

Major Roznenko nods grimly. "In more than theory, I think. With all due respect, Comrade General, the Americans could put such satellites in orbit if they want to. And lasers of very high power could destroy our missiles while they are still in boosting flight, before their rocket engines finish their burn."

"*All* the missiles, Major?" Koshkin asks, a tinge of contempt in his voice. "They could destroy them all?"

Roznenko hesitates, then replies, "No, sir, not all of them. We have more than a thousand, after all, with multiple warheads on most of them."

Koshkin smiles slightly.

"But they could destroy enough of them to make our attack pointless. They could defend themselves quite well, I think."

Aleksandrov says to the younger officer, "Explain that to me, Major. How many could they shoot down?"

"It all depends," Roznenko replies. "As I understand it, the Americans would have to put up a tremendous number of very large satellites — a hundred or more, in orbits of no more than five hundred kilometers. The lasers aboard them would have to be very powerful, many megawatts or more."

"They don't need lasers," Maslov says. "Small pellets

would do just as well, perhaps even better. Buckshot."

"Or protons?" asks Aleksandrov. "I understand you've been working on electrical guns that fire atomic particles . . ."

"Yes, yes, but it's very difficult. Shotguns would be more effective."

"Not over ranges of a thousand kilometers," Roznenko counters. "You need energy beams. They strike with the speed of light. Whatever you can see, you can hit."

"How many of our missiles could be shot down?" Aleksandrov asks again.

Casting a quick glance at General Koshkin, Roznenko answers, "*If* the Americans can put a hundred satellites in the proper orbits, and *if* they can make their lasers work when they have to . . ."

"And if their pointing and tracking systems work perfectly," Maslov mutters.

"Yes, that too," Roznenko admits. "If all that can be done, they can shoot down anywhere from ninety to ninety-nine percent of our missiles."

"Ninety to ninety-nine percent!"

Grechko bangs a chubby fist on his thigh. "Then we must produce more missiles. A thousand more! Five thousand more!"

"And the SALT One treaty, in which we agreed to limit the numbers of missiles?" Aleksandrov asks.

"To hell with it! Our survival is at stake!"

Nikolai Birman, the young economist, raises a slender finger. "Comrades, may I point out something that perhaps you have overlooked?"

Birman is the youngest man of the group, and this is the first time he has been in such exalted company. He has worked for Aleksandrov for several years, rising with him as Aleksandrov has laboriously climbed the ladder

No. 14 Borodin Avenue, Kutusow Prospekt

of the state planning bureaucracy. Birman is lean of body and face, almost ascetic, with hollow cheeks and piercing dark eyes behind his rimless glasses. His hair is glossy black and combed straight back from his high forehead. Aleksandrov has promoted him steadily, despite his being a Jew, and for this Birman is extremely grateful and steadfastly loyal to his boss and patron.

"What is it, Nikolai Maximovich?" Aleksandrov asks. The others all fall silent.

"With all due respect to the military," Birman nods at Koshkin and Roznenko, "and to Comrade Grechko, I think that we must consider the economic facts of the situation. Certainly we could produce a thousand more missiles, or even five thousand, if we had to. But perhaps that is exactly what the Americans expect us to do. Perhaps that is what they *want* us to do."

An amused smile crosses Aleksandrov's face. "Please explain."

Birman gets to his feet, but does not move from the upholstered chair on which he had been sitting. His voice is soft, nearly a whisper. "It is necessary to speak plainly, comrades, without varnishing the truth. As the First Secretary himself has pointed out, our economy is not in good condition. In fact, it is in grave danger."

"We can still build missiles," Grechko mutters.

"Yes, we can," Birman agrees, "but each missile we build warps the economy farther away from where it ought to be."

"It would be necessary to alter the Five Year Plan . . ."

"Drastically," says Birman. "But worse than that, a large change in the missile production schedule will cause even larger changes in every other aspect of the economy. Workers must be taken away from other tasks to build those missiles. They must be fed. And their factories must be fed, too: steel and aluminum and electrical power. Where will it come from? Who will do with less so that we can build more missiles?"

"The Soviet people have made sacrifices before," says General Koshkin.

"Pardon me, sir, but it is not a question of sacrifices. It is a question of the basic strength of the Soviet Union."

"We are the strongest power on earth!" the general snaps.

Birman closes his eyes and bows his head, but only for an instant. His chin comes up and his eyes open again. "Militarily we are strongest, yes. But our economy is weak and it is growing weaker, not stronger."

"Nonsense!"

"Hear him out, Vladimir," Aleksandrov says quietly. "Go on, Nikolai Maximovich."

A look of worry, almost of pain, crosses Birman's face. "Comrades, I detest the necessity of saying so, but the Soviet economy cannot continue along the path it has followed for the past ten years or more. Our nation has vast resources, it is true, but we are consuming those resources faster than they can be replaced. Where once we drew oil from the Caucasus, now we must drill in Siberia, far to the north, where the ground is frozen all year long."

"But we have plenty of oil, and even more natural gas," counters Maslov. "We export trillions of liters of natural gas every year."

"But for how long can we continue to do so?" Birman shoots back. "And the main reason that we export natural gas to Western Europe is to gain hard currency, which we desperately need so that we can buy grain to feed our people!"

General Koshkin glowers at the economist but says nothing. Aleksandrov, leaning against the windowsill with his arms crossed over his chest, is also silent. Major Roznenko,

No. 14 Borodin Avenue, Kutusow Prospekt 213

seeing the general's face redden, decides to speak up.

"Comrade economist," he says, with a smile to show that he is not being hostile, "the Soviet Union finds itself surrounded by enemies — the capitalists of Western Europe, led by the imperialistic Yankees; the yellow hordes of China and Japan; the Moslem fanatics in Iran and the Middle East. The only thing that has kept us from being attacked is our ability to destroy any attacker, including the United States, with hydrogen bombs. The Strategic Rocket Forces are our first line of defense."

"Together with the Red Army," mutters General Koshkin.

"Yes, of course," Roznenko swiftly agrees. Turning back to Birman: "Now then, if the imperialists can develop a weapon that makes our missiles useless, then the whole Soviet Union is wide open to attack. The Yankees, their lackeys in Britain and France, and the Chinese will blow us off the face of the map as soon as they feel confident that we would not be able to strike back at them."

"They won't have to," Birman replies, with quiet intensity. "The Soviet Union is well on the way to self-destruction. The capitalists won't have to bomb us; we are already destroying ourselves."

"That's treasonous!" Grechko thunders.

"Is it treason when a doctor tells a patient he has a cancer," Birman says, facing the fat older man, "and that if he doesn't have it removed he will die? Is that treason?"

"Don't lose your temper, Boris," Aleksandrov says mildly. "I'm sure that Nikolai Maximovich meant no harm by his words."

Casting a grateful glance at his patron, Birman says, "It may be painful to face the facts, comrades, but if we don't face them we are neglecting our duties. It is certainly true that we are surrounded by enemies. But what about the enemies within? Inside the Soviet Union we are beset by enemies, also: low productivity, both in labor and agriculture; outmoded factories; laziness and drunkenness — even in the military. The people are willing to make sacrifices? That's laughable! The people buy whatever they can on the black market. They smuggle Western clothes and books and phonograph records into the Soviet Union. How many of us own videotape recorders that were bought in the West? The people even pilfer state-owned supplies for their own use. They are tired of sacrifices, and every day they grow more tired, less reliable, less willing to wait for the glorious future that the party has promised them."

Before Grechko can say a word, Aleksandrov points a warning finger. "I want you all to remember that what is said in this room tonight remains in this room. Nikolai Maximovich is only saying what the First Secretary himself has warned us about. It may be a hard pill to swallow, but we must face the facts!"

"If we try to build more missiles, to double or triple what we are building today," Birman resumes, "what is to prevent the Americans from putting up more satellites?"

"The satellites are much more expensive than the missiles," Maslov says.

"Even if they are, the Americans have a tremendously rich and powerful economic base to draw from. And they have the economies of Western Europe and Japan, as well."

"Would they really be willing to spend so much on their military?" Aleksandrov asks. "After all, the capitalist nations are more interested in their luxuries. The Americans don't even have a draft . . ."

General Koshkin grumbles, "No, they have a professional army, a full-time army; not the fools and louts we have to conscript for two years and then send home."

Birman smiles wanly at his boss. "To the Americans, spending a hundred billion dollars on satellite weapons is easy. Who gets the money? Their big corporations. It makes jobs for their workers, it builds careers for their military officers, it makes their congressmen and senators feel that they are defending the country — and keeping their factories humming. No soldiers are sent into combat. Nothing happens except that some rockets are launched, and the Americans love rocket launches. The weapons go into space, where nobody can see them."

Maslov, the physicist, passes a hand across his stubbly jaw. "So you are saying that the Americans will outspend us."

"Happily. While we drive our economy deeper into the rut it's already in, the Americans will be covering the sky with satellites."

"Then what can we do?" Maslov asks.

"We will do what we *have* to do," Grechko bellows, "no matter what the price may be."

"But what is that?" Aleksandrov asks. "What is it that we have to do?"

"Let's examine this logically," suggests Major Roznenko.

"By all means," Aleksandrov agrees.

Ticking off the points he makes on his fingers, the major says, "First, our strategic deterrent rests on our ballistic missiles, in their silos and aboard the nuclear submarines. Without those missiles, the imperialists would feel free to attack us. With those missiles, we can destroy any attacker many times over. It is this, and only this, that has kept the imperialists from attacking the Soviet Union."

The others nod their agreement.

"Second, despite all the efforts of our diplomats, the imperialists refuse to negotiate reasonable arms control agreements. They always throw in terms that they know are quite unacceptable, because they don't really want to disarm."

Maslov looks, for a moment, as if he wants to say something to that point. But he lowers his head and remains silent.

"Third — and the reason we are here tonight," Roznenko continues. "The Americans are planning to place into orbit a weapon system that will make our missile forces just about useless."

"It will take them ten years to get those satellites up," General Koshkin points out.

"That doesn't mean we have ten years to make up our minds," says Aleksandrov. The others chuckle, somewhat nervously.

"What are our possible alternatives?" Roznenko asks. "Comrade Grechko was correct when he said we could overwhelm a defensive system simply by producing more missiles than the system can handle. The missiles are cheaper than the satellites, by far. But Comrade Birman says that our Socialist economy is near collapse, and the capitalists can get richer by building more satellites."

Birman shoots an anguished glance at Aleksandrov, who lifts his brows a centimeter as the only sign of reassurance he is willing to give. Birman realizes that he has made enemies this night; he only hopes that Aleksandrov is not one of them.

"Another alternative," Roznenko goes on, "would be for us to build similar defensive satellites so that we can shoot down an American missile attack on us."

"Or a Chinese attack," Maslov says.

"Don't forget the French," General Koshkin adds. "If we should get into a shooting war in Europe, I wouldn't put it past those Frogs to launch their Force de Frappe at us, even if the Americans hesitate to use nuclear weapons."

"*Especially* if the Americans hesitate to use nuclear weapons," Roznenko says, grinning.

No. 14 Borodin Avenue, Kutusow Prospekt

Having recaptured the floor, the major glances at the fingers he has extended, frowns briefly because he's forgotten which number he's up to, and resumes his speech. "But if building more ballistic missiles would damage the economy, as Comrade Birman suggests, then trying to build these huge satellites and their lasers would be even worse. And we're not certain whether the job is technically possible."

"It is feasible," Maslov says. "It is a big job but it can be done. No new breakthroughs are necessary, although the pointing and tracking systems must be made much more accurate than anything done so far."

"So we could put up our own satellites," Aleksandrov says, "if we could afford them."

"We can afford them if we have to," Koshkin says flatly.

"But what if it's a trap," Birman pleads, "an *economic* trap? What could please the imperialists more than seeing us bleed ourselves white over some mad gadgetry in the sky that won't even stop an all-out attack?"

"There is a way," Roznenko insists, raising his voice to still the hubbub that is just beginning to break out, "to counter this American threat without driving ourselves broke. A simple, cheap, effective way that doesn't require any new weapons system or high-speed production lines."

All attention is focused on him.

"Negotiation?" Aleksandrov asks. "We have never been able to get a satisfactory . . ."

Roznenko shakes his head. "No. Not negotiation. Force. We have antisatellite weapons. They are operational. We have been testing them for years, and they work rather well. Each time the Americans put up one of their laser-armed satellites, we can send up one of our ASATs and blow it to smithereens."

The room is absolutely silent for several moments. Aleksandrov studies the faces of the men around him. General Koshkin is nodding slightly. Maslov looks pensive. Birman, pale and uncertain. Grechko's baggy eyes are half-closed. Old Boris will have to be retired soon, Aleksandrov tells himself. But that's another matter.

Aleksandrov breaks the silence. "But if we destroy an American satellite, wouldn't that be a cause for war? Wouldn't the United States attack us immediately then?"

Roznenko hesitates a moment before answering. He has the rapt attention of every man in the room, but the one he wants to impress most is General Koshkin.

Carefully, the major replies, "I was a raw recruit in nineteen sixty-two, when the Cuban missile crisis took place, so I can't tell you what went on at the highest levels of the Council of Ministers and the Central Committee. But we did not go to war then, although we came quite close."

Koshkin stirs uneasily in his chair. "I was in Cuba."

"Really?" Maslov looks surprised and impressed at the same time. Aleksandrov knew it, of course. And so did Roznenko.

Bitterly, Koshkin says, "I watched our construction troops build the missile launching pads. And I watched them tear them down again."

"The point I want to make," Roznenko says, "is that we have come very close to war with America on several occasions. Neither they nor we have unleashed our nuclear missiles. I don't believe the Americans would go to war over their satellites. Not if we orchestrated our response properly."

Aleksandrov says slowly, "You know, there are many Americans who are against this entire idea. They want to keep space pure, no weapons."

Koshkin laughs. "Too late for that, by nearly twenty years."

"Destroy the American satellites as they are put into

orbit," Maslov muses. "That's a dangerous game. An *extremely* dangerous game."

"What alternatives do we have, comrade?" asks Roznenko. "Allow the imperialists to get the upper hand over us? Bleed ourselves white, as Comrade Birman put it, trying to match them? No. We must take the risk. Show the Yankees that we will fight to defend ourselves. By blowing up their satellites, we will convince them that we are ready to strike them, their cities, if need be. They'll back down."

"But if they don't?" Maslov asks, in a fear-hollowed voice.

Aleksandrov, heading for the locked liquor cabinet across the room, quips, "If they don't, then our worries will be over, in approximately half an hour!"

13

The Russian Decision

VIKTOR ILYICH ALEKSANDROV is a fictitious character, but his position and problems are based on careful research, interviews, and personal acquaintance with scientists, artists, bureaucrats, and politicians on both sides of the Iron Curtain.

American perceptions of the policies of Soviet Russia tend to run toward the extremes: there are doves and there are hawks, with the former usually identified as liberals in domestic politics and the latter as conservatives. Although these simplifications distort the actual situation to a considerable degree, they provide useful labels.

The liberal, dovish position is that the Russians are not devils intent on conquering and enslaving the world; they are human beings saddled with a repressive government, attacked since time immemorial by foreign invaders, and burdened with a backward and unproductive economic system.

The conservative hawks believe that the Kremlin seeks world revolution, that the Soviet leaders will be satisfied with nothing less than the destruction of all capitalist nations, especially the United States. To achieve these ends, the Russians maintain the world's largest army, build the heaviest missiles, and support revolutionary regimes and terrorist organizations around the globe.

To Viktor Ilyich Aleksandrov, however, the world situation looks quite a bit different.

Aleksandrov is a realist, or at least he tries to be. He is a capable, talented politician who has risen almost to the highest rank in the Soviet government and the Communist Party. (Although the government and the party are not synonymous in the USSR, no one gets into the government without being a party member in good standing.)

Aleksandrov was born in 1929, in Moscow, of Muscovite parents. Among Russians, that is the equivalent of being born in New York City, of Manhattanites. His father was an electrical engineer, killed in the defense of the city in 1941. He and his mother were moved to one of the industrial towns established behind the Ural Mountains, where she worked as a clerk in a tank factory that, as late as 1943, had no roof and no heating system. Viktor Ilyich lived in a tent with his mother through two bitter winters. When he was fifteen he tried to enlist in the Red Army but was walked back to his one-room apartment in a new wooden frame two-story house by the recruitment sergeant, who admired the youngster's zeal but advised him to wait a few more years and add a few pounds to his bones.

By the time he was big enough to enlist, the war was over and Viktor was the local leader of the Young Communist League. He was drafted, as every Soviet male citizen was, and worked his compulsory two years at helping to clear the rubble from the streets of Kharkov and get the public transportation system running again. One of his most vivid memories, even to this day, is his first sight of that devastated city, bombed, shelled, fought over street by street. Hardly a building still had its roof or all four walls standing. Bullet scars "decorated" homes and factories, churches and theaters alike. For months they were still digging bodies out of the ruins.

To Aleksandrov, and all the Russians of his generation, the twenty million Russians killed in World War II are not merely a statistic: they are fathers, mothers, sisters, wives, brothers, husbands, sons, and daughters. By contrast, slightly more than two million Americans have been killed in *all* the wars the United States has fought since 1775.

From his level in the Soviet government, Aleksandrov is constantly made aware of two salient facts. First, the USSR is surrounded by hostile nations, from the capitalists of Western Europe and their American overlords to the deviationist Chinese and their population of more than one billion. The Americans are implacably hostile to the Soviet Union and to the very idea of Communism. Perhaps the American people, the exploited workers and farmers, would be less inimical to Russia and the inevitable triumph of Communism if they were not so misguided by their government and the capitalistic press. But with the single exception of Franklin Roosevelt during the years of World War II, Washington has consistently shown nothing but hatred for Soviet Russia. Reagan has been blunter than even the irascible Truman, calling Russia the source of all the evil in the world. His former National Security Adviser, that frontier judge William Clark, strutting in his cowboy boots and ten-gallon hat, echoed his chief's sentiment: "The Soviet state is a bizarre and evil episode of history whose last pages are even now being written." Hatred, pure and simple.

The second salient fact that worries Aleksandrov is that the Soviet economy, which has performed miracles of production all his life, is faltering badly. Somehow, even though production of steel, coal, petroleum, and other basics remains among the highest in the world, the nation's gross national product and income per capita are dwindling. Soviet Russia is becoming poorer every year, and the people are becoming more restless, more openly critical of the government, and less reliable.

In 1946, when he was shoveling rubble from the shattered streets of Kharkov, Aleksandrov thought that the era of Communism was near at hand. The war against the fascists had been won, at terrible cost, of course, but won. The Red Army controlled Eastern Europe. Western Europe was devastated, smashed by American and British bombing raids and the battles that raged from Normandy to the Elbe River. It was only a matter of time before Italy, France, and the rest of the continent turned to Communism.

But it never happened that way. Somehow, the only nations to remain within the Russian sphere were those on which Soviet soldiers were already firmly camped. The Red Army even retreated from its share of little Austria, eventually, in return for having Great Britain, France, and the United States pull back too and let Austria become a neutral.

The Americans had the atomic bomb, of course. They thought that its awesome destructive power made them invincible, and loudly threatened to use the bomb on Russia unless the Soviet government knuckled under to them. The imperialists did not reckon on Joseph Stalin's iron will nor on the determination of the Soviet people. Aleksandrov remembers those postwar days with some pride. Battered but victorious, its richest and most productive areas reduced to smoking rubble, with twenty million dead to bury, the Soviet people still pulled themselves out of the rubble to become the world's most powerful nation.

Stalin drove them ruthlessly, of course. Like Ivan the Terrible and other great Russian leaders, he went a little mad toward the end. But there was so much to do, so much lost ground to make up for, and so little time. In less than five years, Soviet scientists exploded their first atomic bomb, and then went on to produce a hydrogen bomb *before* the Americans did. And while the Americans were building bigger and bigger bombers to carry their nuclear attack to Soviet soil, Nikita Khrushchev outflanked them by building missiles that could deliver hydrogen bombs on American cities in only thirty minutes from launch.

That staggered the Americans. They had been able to rebuild Western Europe and Japan while Eisenhower and the Dulles brothers were establishing a ring of alliances around the Soviet Union: NATO, SEATO, and others. But the missiles leaped across such defenses and threatened the American homeland, just as the imperialists threatened Russian homes and cities.

Sputnik was a huge success. It demonstrated to the world that Soviet rockets were real, that the progressive Socialist peoples were lifting their vision to the stars, and that their technology was strong enough to defend them against any aggressor. When the capitalists fomented their counter-revolutionary uprising in Hungary in 1956, canny old Nikita made it clear to all the leaders in Europe that if they should try to intervene, Soviet rockets would rain down on their cities instantly.

The American reaction was predictable, Aleksandrov thought, with the benefit of perfect hindsight. Stunned by the success of Soviet rocketry, the Yankees invented a mythical "missile gap" and escalated the arms race. Not content with the bombers and military bases they had placed all around the Soviet Union like a ring of bayonets pointing at Moscow, they started frantically building missiles, as well. Khrushchev tried to surprise them by placing missiles in Cuba to give them a taste of their own medicine.

But they found out about the missiles before they became operational and forced Khrushchev into a humiliating backing-down.

That cost Nikita his job, but at least he was allowed to retire peaceably. There was no need to put tanks in the streets and shoot him, as had been done with Beria, after Stalin's death.

To Aleksandrov, the Brezhnev years were a time of constant, sometimes rapid, advancement, not only for him personally but for the Soviet Union as a world power. The Americans twisted and turned, got themselves mired down in Vietnam, exposed their dirty Watergate linen in public, and were forced to accept the idea that they could never — ever — again be militarily superior to the Soviet Union. Aleksandrov believed that *this* was the real reason for the collapse of morale and discipline in the United States. It finally became clear to the American people that they could no longer feel superior to the Russians. In fact, in some of the key areas of power, the USSR was significantly ahead of the Yankees.

But the higher he rose in the government and the party, the more Aleksandrov became aware of the internal weaknesses of his nation. A lesser man might have become totally disillusioned with the ineptitude, the corruption, the cynicism, of the bureaucrats surrounding him. Worse still was the growing odor of stagnation rising from every part of the country, even the military. It reminded Aleksandrov of his first whiff of Kharkov after the war: the odor of lingering death.

It was easy for fat old Grechko and General Koshkin to say that the Soviet people would endure the sacrifices necessary to defend the nation. They lived in swank apartments, as Aleksandrov himself did, and shopped in special stores reserved for the *nomenklatura,* the elite government and party members who were too high up, too important, to stand in line all day waiting to buy a couple of lamb chops or a washer for a dripping faucet. But the economy was faltering, there was no doubt of it. Even in the areas where the USSR led the world, steel and oil production, the rate of increase was slowing down dangerously. And no matter how much the factories and mills and mines seemed to produce, there were always shortages: not enough grain, not enough automobiles, not enough houses. The people were becoming just as corrupt and cynical as the worst of the bureaucrats. The black market, the "second economy" that everyone knew about, soaked up more and more of the people's time and effort and cash.

Productivity was the key. In America, less than 3 percent of the people were farmers, yet they produced enough food to keep the Americans fat and to have a surplus almost every year. Yankee grain was sold all over the world — even in Soviet Russia, where more than 20 percent of the population worked on the farms, and produced nothing but shortages.

The Americans. Aleksandrov pondered long and hard over them. If only they would take off some of the pressure, if only they would admit to themselves that the USSR is a world power worthy of their respect, perhaps even their admiration. But no, the prevailing American attitude was that the Soviet government ruled a nation of slaves who would someday rebel and overthrow the Communist Party. The Yankees actually dreamed, in their heart of hearts, that Russia would return to capitalism someday, like a repentant little boy who had done something naughty and then asked for forgiveness. Worse than that, the imperialists also hoped that the Soviet Union would break apart, that the republics of Georgia, the Ukraine, and the so-called Moslem states would separate from Mother Russia and form their own independent nations. They published magazine articles and books showing that the non-Russian parts of the Soviet Union were growing in population faster than the Slavic states in general and faster than true Russia — the Russian Soviet Federated Socialist Republic — itself.

They would like that, the capitalists would: seeing Russia drown in a tide of Moslem and Asian hordes. But it will never happen, Aleksandrov told himself. The USSR had been united by Lenin and would remain united despite anything the capitalists did.

At the end of the Great Patriotic War, after Soviet troops had born the brunt of the fighting, after the Americans and British had delayed their Second Front beyond all conscionable excuses, after the Red Army had taken Berlin, the Americans had immediately stopped all aid to the USSR. Even before the war with Japan was concluded, they stopped their Liberty ships in midocean and turned them away from the Soviet Union. It was as if the capitalists were afraid that their trucks or engines or grain would somehow become Communist once the Nazis had been annihilated.

They crowed over their atomic bomb, which they had developed in secret. The British knew about it, but the Americans did not tell the USSR until Truman made a veiled reference to Stalin about it at the Potsdam Conference, just a few days before Hiroshima. Fortunately, Aleksandrov thought, Russian physicists already knew the basic science involved and were able to overtake the Yankees with a few years of hard work.

When Khrushchev pushed the scientists to develop missiles that would outflank the imperialists' ring of military bases drawn around the Soviet Union, and then drew world attention to their success by using the same rockets to launch Sputnik and the other space probes, the world was stunned. Suddenly the capitalists were forced to admit that Soviet science and technology were not merely the equal of the West's; in some respects they were better. But did the Americans accept this accomplishment and welcome the USSR into the rank of technological superpowers? Certainly not. Aleksandrov remembered the dismay and disappointments as the Americans worked themselves into a lather over the "missile gap" and the "moon race." Kennedy came into the White House talking tough, and the Americans began producing missiles so fast that there were real fears in the Kremlin of an imminent war. And at the same time the Americans started a completely new line of rockets that eventually carried their astronauts to the moon.

Aleksandrov had not been the only Russian staggered by the miracle of American productivity. Strangely, though, most of the Americans themselves failed to appreciate what they had accomplished. They nearly killed their space program during the 1970s, just when they could have established a stranglehold on space.

Leonid Brezhnev was not as ruthless as Stalin or as flamboyant as Khrushchev, but he was the right man at the right time, as far as Aleksandrov was concerned. Under his steady, firm direction, Soviet military power gradually but inexorably gained parity with the West. When the Czechs became obviously counterrevolutionary, Brezhnev coordinated a brilliant set of military and political maneuvers involving all the Warsaw Pact nations and brought Czechoslovakia securely back into the Socialist camp. And when the generals were ready to invade Poland in 1981, Brezhnev not only stayed their hand, but actually sacked the hottest heads among them and strengthened his own control of the military.

Yet, though Soviet military might had at last reached a par with the Americans', no Russian could feel totally secure, Aleksandrov knew. The Americans still had a nuclear arsenal that could destroy the USSR at a stroke, and they were determined to place new missiles with nuclear warheads in Europe. All the long years of patient work and propaganda, the Peace Movement, the Nuclear Freeze Movement, none of it deterred the imperialists in the slightest. They quite openly declared that they would use nuclear weapons against the Warsaw Pact forces if war broke out in Europe. They even developed the neutron bomb, which would minimize blast damage to buildings while maximizing radiation output to kill people; Aleksandrov thought of it as the ultimate capitalist weapon. And they gave as their excuse Soviet Russia's deployment of new missiles to replace the aging and obsolete medium-range missiles that had been in place for twenty years.

That was bad enough, but this idea of building space satellites to shoot down missiles was a far more serious problem. It struck at the root of Soviet power. In its way, Reagan's proposal to build such laser-armed satellites was just as clever a flanking movement as Khrushchev's original move to ballistic missiles.

Reagan came into the White House talking like the sheriff in a cowboy movie. It reminded Aleksandrov of Kennedy's blazing rhetoric, twenty years earlier. Yet the capitalists were devious, and certainly could not be trusted to back their words with actions. Kennedy eventually signed the Nuclear Test Ban Treaty with Khrushchev — but only after the Cuban missile crisis had brought them to the brink of war. And the toughest-talking Yankee of them all, Nixon, was the President who agreed to détente, even while he escalated the fighting in Vietnam. The Americans were very tricky, Aleksandrov realized.

Perhaps Reagan's tough talk was a bluff, a front to mask the essential American weaknesses. But if so, was Reagan merely stalling for time while the Americans re-armed, or would he be genuinely willing to negotiate a new form of détente? Aleksandrov thought that re-arming was much more likely than negotiation. Birman was right, the American military-industrial complex welcomed re-armament. Already their laboratories and factories were humming with new orders for weapons. The progressive elements in the Congress were resisting the huge increases in defense spending, but they had been unable to halt the new B-1 bomber or the Trident submarine or even the highly dangerous MX missile. Aleksandrov uneasily watched the reports of the American space shuttle, knowing that it was essentially a military vehicle, built by NASA to specifications laid down by the U.S. Air Force. The Soviet shuttle would not be ready for its first flight tests for several years, although the Strategic Rocket Forces had established their functioning space station, Salyut 7, at long last.

The Yankee press called Reagan's orbital defense system a Star Wars scheme. Were they genuinely derisive, or was their sarcasm merely a clever ploy to deflect serious consideration of the idea by Soviet planners? Aleksandrov neither knew nor cared. He had been ordered by the First Secretary's most trusted aide to examine the problem in depth, from every possible angle, and to come up with a recommendation that could be taken to the First Secretary himself.

The meeting in his apartment that night several weeks earlier had been Aleksandrov's first step in educating himself. He had worked hard and intensely since then, and though he did not consider himself an expert on space technology, at least now he understood the essentials of the situation. He was beyond the confusion he had first felt when suddenly confronted with an ocean of such acronyms as BMD (ballistic missile defense), ABM (antiballistic missile), EMP (electromagnetic pulse), HEL (high-energy laser), MAD (Mutual Assured Destruction), DEW (directed-energy weapon), and so on.

The more he studied the situation, the more serious

the problem appeared to be. If the Americans could establish space platforms carrying weapons that could destroy Soviet missiles, then the underlying base of Soviet power was in danger. Once the Americans felt secure in the knowledge that their homeland could be protected from a nuclear attack, what was to stop them from launching a devastating attack on the Soviet homeland? Aleksandrov could see the sequence of events in his mind's eye:

The capitalists would foment a crisis somewhere, anywhere. The Middle East, Southeast Asia, Central America; there were plenty of hot spots on the map. Knowing that Soviet missiles were virtually useless, they would bully the USSR into accepting their terms, speaking softly perhaps, but carrying the big stick of instant nuclear annihilation. If the Soviet Union backed down and allowed the imperialists to have their way, the Americans would come back with a new crisis, new demands, new ultimatums. Nothing would satisfy them until they had destroyed Communism entirely and turned the USSR over to a counterrevolutionary ruling clique. Eventually, inevitably, the Soviet people would fight. They would fight to defend their rights, their way of life, their government. And the American leaders in the Pentagon, rubbing their hands with glee, would launch their missiles and bombers against their hated foe. As soon as their missiles rose from their silos and submarines, the Soviet Union would retaliate. But while the Soviet missiles were being destroyed by American satellites armed with lasers, the Yankee missiles would fly, unimpeded, to their targets and obliterate every city in Russia. There would be twenty million dead within a few minutes, rather than in four years, as in World War II. And far more would die of radiation, starvation, and disease over the following weeks. There would be no more Soviet Union. There would be no more Russia.

Once he had framed that picture of overwhelming destruction and death in his mind, Aleksandrov came to his first and primary conclusion: Koshkin had been right, there is no price too high to pay for survival. The threat of orbiting ABM weapons had to be countered, no matter how much Birman worried about what it would do to the Soviet economy. Perhaps, Aleksandrov told himself, a massive armaments program like this would boost the economy. After all, the military sector of the economy was apparently the only one that worked consistently well.

Like a man who has put all fears behind him, Aleksandrov squared his shoulders and began drafting a detailed memorandum for the First Secretary's perusal. He carefully reviewed the military situation that the Star Wars idea had created, and then — with Maslov's help — added a little primer on the scientific facts. But the bulk of his memo dealt with his recommendations for action.

There are four possible courses of action, he wrote: attempt to prevent the Americans from establishing ABM satellites in space; attempt to match them with similar Soviet satellites; attempt to negotiate an agreement that would produce a solution satisfactory to the Soviet Union. The fourth course was Grechko's solution: simply build enough additional missiles to overwhelm the enemy's defenses. But how many would be enough? The uncertainties were very high.

In the first category, the methods available to prevent the Americans from placing ABM satellites in orbit range from propaganda to ASAT attacks on the satellites themselves. Certainly worldwide opinion could be marshaled against the idea of placing weaponry in space. The provisions of existing treaty agreements, such as the 1967 Outer Space Treaty, the SALT I agreements, and the 1972 ABM Treaty, could be used as bases for arguing that ABM satellites are violations of international law. Popular movements against space weaponry could be mounted, similar to the

Nuclear Freeze Movement. The United Nations could be used as a forum to educate the masses about space weapons and to pressure the United States government against development or deployment of such satellites.

There are sizable elements within the American government and among the American people (Aleksandrov continued) that are opposed to the very idea of placing weapons in space. It is imperative to secure their support in any anti–space weapon campaign. During such a campaign, it is very important that the Soviet Union refrain from creating the impression that it has placed military satellites in space. Launches of nuclear-powered surveillance satellites should be restricted so that embarrassing incidents like the crashes of Kosmos 1402 in 1983 and Kosmos 954 in 1979 are not repeated during this time. Also, tests of antisatellite weapons should be curtailed for the duration of the campaign if they cannot be kept secret.

Aleksandrov then recommended that if the Americans go ahead and begin placing ABM satellites in orbit anyway, then the Soviet Union will have to consider very seriously the option of using its ASAT weapons to destroy them. Great care should be taken to destroy only the satellites themselves, without injuring or killing American astronauts or damaging their space shuttle vehicle. It is likely that destruction of the satellites would force the Americans to the conference table, but endangering the lives of their astronauts could be an excuse for them to start a war. He added that the Americans might take reprisals for any Soviet attack on an ABM satellite. The United States is developing an ABM missile, Aleksandrov pointed out, launched from a high-flying F-15 fighter aircraft. If the Soviet Union destroys an American ABM satellite, the United States may well retaliate by destroying selected Soviet satellites. Its most probable targets would be surveillance satellites, which orbit at relatively low altitudes. Aleksandrov emphasized his opinion that the Americans would be very unlikely to attack Salyut 7 or any other Soviet manned spacecraft, fearing that such an attack on cosmonauts would quickly escalate into all-out war.

The second possibility is for the Soviet Union to develop its own ABM satellites, armed either with lasers or with particle-beam weapons.

Aleksandrov recommended that very high priority be given to development and testing of the necessary equipment. The existing Salyut 7 space station is an excellent platform from which to test energy-beam weapons in space, he wrote. It is already in a low orbit and is manned by two or more cosmonauts almost constantly. Since early 1983, when the second unit was mated to it, Salyut 7 has been large enough to house six cosmonauts or even one or two more (if they are in love, the space engineers joke). The space station is resupplied by unmanned Progress spacecraft every week; it would be relatively simple to assemble the hardware and personnel for a test of an energy-beam weapon in orbit.

If the tests prove satisfactory, then the Council of Ministers and the Central Committee must decide whether or not the USSR should begin a full-scale program to place a hundred or more energy-beam-armed satellites in orbit. The costs would be enormous, Aleksandrov warned. But if such weaponry could negate the American, British, French, and Chinese missiles aimed at Russia, the prize would be well worth the cost. The first few such satellites could be disguised as new Salyut space stations, Aleksandrov suggested, allowing the USSR to place at least a few of them into orbit before the Americans realize what is being done.

There would be intermediary benefits from having such weapons in space, he added, even if only a few are placed in orbit. First, the energy-beam weapons would be able to attack other satellites, as well as ballistic missiles. Currently, the Soviet ASAT weapon, which is little more than a small packet of explosives launched into orbit by a modified SS-9 ICBM, places its warhead into an orbit that brings it close to the target satellite and then detonates the explosive, destroying both the killer satellite and its target. This system can reach target satellites only in relatively low orbits. A powerful energy-beam weapon, however, fired from space, could most likely destroy or at least disable satellites in almost any orbit. Aleksandrov noted that more than 80 percent of the American military's worldwide communications are carried by a few communications satellites in geosynchronous orbit, 22,300 miles above the equator. The Soviet Union's existing ASAT system cannot reach them. But a space-born energy-beam weapon could.

Moreover, if the USSR can get a few beam weapon satellites into orbit before the Americans, the weapons might be used to demolish any ABM satellites that the Americans attempt to place in space. It might be, Aleksandrov suggested, that the nation that is *first* in putting such weapons in orbit would achieve a decisive military advantage, one that would give it dominance not only in space, but on the ground as well.

Aleksandrov carefully pointed out that energy weapons fired from satellites probably would not be able to destroy aircraft, such as the American B-1 bomber or cruise missiles, unless they fly very high in the stratosphere, where the air is so thin that it offers little interference to an energy beam. Low-flying aircraft and targets on the ground or at sea would be safe from energy beams fired from space — for the foreseeable future. He added, though, that the development of this space weaponry could pay dividends in weapons development for the Red Army and Navy. High-power lasers fired from the ground or from the decks of ships could provide excellent antiaircraft protection, he suggested, and might offer a counter to the threat of American cruise missiles.

Finally, Aleksandrov wrote, it might be possible to prevent the Americans from deploying energy weapons in space by negotiating a treaty with them that bans all weapons from space. Already, the Nuclear Test Ban and 1967 Outer Space treaties have outlawed testing or basing nuclear weapons in space. The moon has been demilitarized. The United Nations–sponsored Moon Treaty even prohibits private corporations from working in space, unless authorized by a U.N.–controlled governing body. (Aleksandrov quickly pointed out that the United States has not signed the Moon Treaty and therefore is not bound by its restrictions. But international pressure may induce the Americans to sign it eventually.)

A treaty banning all weapons in space would be extremely difficult to negotiate, Aleksandrov noted. The Americans will insist on inspection of Soviet space payloads. The USSR, on the other hand, has already defined the American space shuttle as a weapons system, since it is capable of carrying military payloads into orbit. This definition would have to be modified once the Soviet space shuttle has its first successful flight, he wrote. The Foreign Office should be instructed to work with the Strategic Rocket Forces to see whether it is possible to create wording for a treaty that will outlaw the West's weapons in space while still providing the necessary loopholes for the Soviet Union to place energy weapons in space.

In his final paragraph, Aleksandrov wrote:

> The new technology of energy weapons has created a new situation in the strategic sphere. Since the first atomic bombs were developed, strategic planners have concentrated on the offense. Atomic (and later hydrogen) bombs were so powerful, and the missiles to deliver them were so swift, that no defense against them seemed likely. Now, energy weapons and other new technol-

The Russian Decision

ogy promise to provide a possible defense against nuclear missiles. Soviet scientists and engineers, who have led the world in the development of rocketry and space flight, have the capability of developing the weapons that can shield the USSR from nuclear attack. Although the political and economic consequences of such a development are complex, one fact stands out clearly: *the Soviet Union cannot afford to be second to the West in the development and deployment of energy-beam weapons in space.*

14

19, St. Mark's Crescent, Regent's Park

THE HOUSE is comfortable, though far from elegant. It is one of the row of houses set along the curving, quiet street. From its living room you can look right through its narrow length and out to the canal and towpath running behind the houses. Small boats are tied up at the docks behind each house. You can travel through much of London along those canals; Britain is crisscrossed by nearly five thousand miles of inland waterways, built before the railroads.

Colonel Mitchell Stanley is in mufti, his Air Force uniform replaced by a casual, rather conservative grayish-blue pinstripe suit that he splurged for on Savile Row.

"Do you live here?" he asks.

David Green, a fellow American, grins and shakes his head. "Nope. Friend of a friend lent me the place for the month. Usually when I'm in London I stay in a hotel. This is a lot nicer, though."

Green is at the bar, which serves as a room divider between the living room and kitchen, busily mixing a pitcher of martinis. He is a compact, wiry man in his late fifties,

19, St. Mark's Crescent, Regent's Park

unmistakably Californian in his light sports jacket and slacks, open-necked shirt, and his tennis player's tan. Colonel Stanley, nearly a head taller, twenty pounds heavier, and with a drooping mustache, has the introspective, almost melancholy look of a world-weary intellectual. At parties, Londoners mistake him for a college professor; they usually find it hard to believe that he is one of the founding members of the U.S. Air Force Space Command.

"Lord Mattingly and the others should be here in a couple of minutes," Green says cheerfully, stirring the pitcher with vigor. "I understand the old man likes martinis. Gin martinis."

Stanley's nose wrinkles slightly. The tumbler in his right hand holds nothing stronger than Chablis. Green could not find wine glasses in the bar or in the kitchen cabinets.

David Green is an engineer by training, a salesman by job description, and a diplomat by nature. The company he works for, a giant aerospace corporation that builds military and commercial jet aircraft, missiles, rocket boosters, and many of the components of the space shuttle, sends him to Britain and the continent at least twice a year. On his current mission he is not attempting to sell hardware; his task this time is to determine how key NATO officials are reacting to President Reagan's proposal for building satellites to defend against ballistic missiles.

The doorbell rings and Green hurries to answer it. Stanley hovers in the doorway connecting the living room with the front hall. He sees three men welcomed into the house by Green; the oldest one is obviously Lord Mattingly. The other two are quickly introduced as Dr. Rolf Gunnerson, a Danish physicist serving as a scientific adviser to NATO, and Louis Arlon, who is Flemish and a high-ranking intelligence analyst in the Belgian government.

Colonel Stanley has served in a dozen foreign nations, but this is the first time he has been in London. One look at the Viscount Mattingly, and he immediately thinks of Winston Churchill. The viscount is smallish and slightly bent, but his face is weathered like an old sea dog's. His handshake is firm and his growling "Good evening, Colonel," fills Stanley's head with the legends he has heard all his life about the Battle of Britain and "their finest hour."

Green ushers the little group into the living room, skipping about like a playful puppy, offering drinks, cigars, and the comfortable chairs scattered around the dark fireplace. Mattingly accepts a martini and sinks into the big wing chair facing the hearth.

"Too warm for a fire," he grumbles. "Still, it's always comforting to watch the flames, don't you agree?"

"Oh, yes, indeed," Green quickly replies. "I could get one started . . ."

"No need," Mattingly holds up an imperial hand.

His chair becomes the focal point of the room as the others array themselves on either side of him. Gunnerson, tall and rawboned, wearing tweeds, and puffing on a pipe that is worn and stained with age, sits alongside Arlon. The Belgian is a slight, darkish man in his forties, with deep, probing eyes. The two Americans arrange themselves on the other side of the viscount, closer to the bar.

"Mr. Green has asked us to convene here this evening," Lord Mattingly states, "to sound out our opinions on this scheme to place satellites in space that can shoot down ICBMs. What do they call it, David, 'Star Trek'?"

"No, sir. Star *Wars*."

Mattingly nods and takes another sip of his martini.

"I'd like to know more about it myself," says Dr. Gunnerson. "I'm not sure that it can be done."

"We've gone through the physics," Colonel Stanley replies, in a quiet voice. "It can be done."

"What about the kill mechanics? Can a laser beam actually destroy a missile? Fast enough to be of any practical use?"

"It's a matter of power density on target," Stanley says, "which in turn is a function of power output and optics. I could go through the equations with you, but I'm afraid it would bore the rest of us."

"You're a physicist?" Gunnerson asks.

"I have a degree in physics, yes, sir," says the colonel.

Mattingly regains the floor. "Granted that one laser in space can shoot down one missile. How many lasers would be needed to stop a full-scale attack? Ten? A thousand? How much would they cost? How vulnerable would they be?"

"Extremely vulnerable, I should think," says Louis Arlon. "They must be huge, these laser satellites. They will make very easy targets for Russian antisatellite weapons."

"They could defend themselves," Stanley argues.

"After they are completely assembled," counters Arlon, "not before. And even then, their fuels could be exhausted by fending off antisatellite weapons *before* the Russians launch their missile attack."

"The lasers themselves might stay on the ground," Stanley says, "and bounce their beams off mirrors placed in orbit."

"The mirrors will be vulnerable then," counters Arlon.

"That's not a likely scenario, is it?" asks Green. "I mean, after all, if the Soviets start attacking our satellites, even if they're just mirrors, that would be a tip-off that they're getting ready to launch a full-scale attack at us, wouldn't it?"

"Not necessarily. It could simply mean that they don't want to have your weapons in orbit over their heads," Arlon says.

"How many missiles would it take to knock out one of these satellites?" asks Gunnerson. "How many boosters do the Russians have?"

Mattingly points a finger at Colonel Stanley. "What would you do if the Russkies started to build such satellites of their own?"

"I guess I'd try to prevent them from completing them," Stanley answers. "I'd use our ASAT against them, if we couldn't get them to stop any other way."

"What's your feeling about the idea, Lord Mattingly?" asks Green.

Mattingly is a Welshman who has spent his entire life in service to the Crown. He volunteered to gather intelligence in Spain during the civil war there in the late thirties, and was wounded in the spine by a German bullet. He spent World War II on the decks of British warships, fighting the Battle of the Atlantic against U-boats and icy storms. After the war he became one of those *éminences grises* who directed the course of the Royal Navy without being hampered by public scrutiny. Officially retired, he still exerts his considerable influence both in Whitehall and the House of Lords.

"It sounds like science fiction to me," Mattingly replies, without hesitation. Then he adds, "But so much of today's world seems like science fiction that I suppose we must consider the idea quite carefully."

"It's a revolutionary idea," Colonel Stanley says. "For the first time since the development of the atomic bomb, we have a chance to produce a defense against nuclear attack."

"I wonder if that's a good thing," Gunnerson says. "I mean, it's terrible to live under the threat of nuclear war, I agree, but it's been that same threat that's kept the superpowers from going to war."

"Indeed," Mattingly agrees. "When you think that the

immediate cause precipitating World War One was the assassination of an archduke, and the trigger for World War Two was a crisis over the city of Danzig, we have had *excellent* reasons for going to war against the Soviets: the Berlin crisis of nineteen forty-eight, the Korean War, the Hungarian uprising in 'fifty-six, the Berlin Wall in 'sixty-one, the Czechs in 'sixty-eight, Vietnam, the Poles — wonderful reasons for starting a global conflagration!"

"You forgot the Cuban missile crisis," Green interjects. "We damn' near did go to war over that one."

Mattingly mutters a grudging acknowledgment.

"Yes, but both Kennedy and Khrushchev backed away from war," Gunnerson says. "And they negotiated the Test Ban Treaty afterward."

"And the Outer Space Treaty, which prohibits placing nuclear weapons in space," Stanley adds.

" 'Weapons of mass destruction,' " Lord Mattingly says. "That is the wording of the treaty. I presume that lasers and such are not considered to be weapons of mass destruction."

Green answers, "No, they are weapons of pinpoint destruction."

"What about the SALT One agreements? Didn't you chaps sign a treaty with the Russians that prohibits building ABM systems?"

"The lawyers are arguing over that one," Green admits. "The treaty is worded loosely enough so that we can at least develop and test an orbital ABM system. If it works, and it looks like we want to go ahead and deploy the full system, we'll either find a loophole in the treaty or abrogate it."

"The Russian propaganda machine would have a field day over that."

"Let 'em! I'd rather be protected from Russian missiles than Russian invective."

Mattingly chuckles, a coughing, grunting sound from deep in his chest, like the rumble of a lion.

"Have you considered," Arlon asks, staring at the two Americans, "that your Star Wars defense could mean the end of NATO and the loss of Western Europe to the Soviets?"

Green blinks with surprise. Stanley, frowning slightly, asks, "How do you come to that conclusion?"

"Consider," Arlon says, in English that is only slightly accented. "America spends billions of dollars to place these satellites in orbit. Hundreds of billions, no doubt. It takes years, but the task is completed. The system works. America is now protected against Russian missile attack."

"So is Europe," Stanley says.

"Yes, perhaps so. But the important fact is that America now feels safe against the Russian missiles. Why bother to keep troops in Western Europe? Why bother with the defense of the NATO nations at all? America is now safe — just as she was when she had two vast oceans protecting her from European and Asian wars. America can return to her old isolationist ways and let Europe fend for itself."

"If that happened, it would be Europe's own fault," Stanley says.

"How so?"

"For thirty years, no, more like forty, the United States has stood guard against a Russian invasion of Western Europe. Side by side with our NATO allies, we've put our troops and planes on the line. We've let the Soviets know that any attack on a NATO nation would be considered as an attack on the United States."

"Commendable," says Arlon, with a smile that is close to a smirk.

"During all those decades, though," Stanley continues, "the NATO nations have never exerted themselves to maintain enough defensive strength to counter the Warsaw

Pact threat. The nations of Western Europe have never spent the money or the effort necessary to defend their own territories properly. They've depended on Uncle Sam to do the job for them."

"And why not?" asks Arlon. "Uncle Sam has the nuclear weapons. As we said only a few moments ago, it is your nuclear bombs that keep the Russians from attacking, not our tanks and infantry."

"Yeah. Western Europe has arranged things so that if the Russians want to attack the NATO countries, they'll be forced into a nuclear war with the United States."

"Precisely."

"It's worked out pretty good for you," Stanley continues. "You spend as little as you can on your military, and then ask the U.S. to shield you with nuclear weapons because the Russians are so much stronger than NATO."

Mattingly replies, "Surely you don't expect the NATO nations to match the Warsaw Pact division for division, tank for tank. That would be political suicide for the leader of any NATO country even to suggest it. The people would never stand for it."

Stanley says, "So you put yourselves in the position where you need battlefield nukes to make up the difference in conventional strength. And you know that the Russians have their own battlefield nukes. And if they use them here in Europe, the United States is obligated to attack the homeland of the Soviet Union."

"Colonel Stanley," says Mattingly, "those battlefield nuclear weapons are *political* weapons. They are in place to remind the Russians that if they start a war in Europe, their own homeland will be destroyed."

"And so will ours," Green says.

"All because the people of Western Europe aren't willing to do the work necessary to keep their own defenses up to the level needed," says Stanley.

Green adds, "What gets me is that when we try to modernize the tactical nuclear force that the West Germans and other NATO countries asked us for, the goddamned people start rioting in the streets and demanding that we get the hell out of Europe altogether."

"We don't want to be a nuclear battlefield any more than you do," says Gunnerson.

"Europe has seen enough battles," Arlon agrees. "My own land has been devastated twice in this century by wars."

"I shall always think of Belgium," mutters Mattingly, "as the country in which Waterloo is situated."

"Bastogne," says Green. "I was only a kid then, but I remember Bastogne and the Battle of the Bulge."

"What is nostalgia to you," Arlon says sharply, "is painful memory to us. My grandfather was shot by Prussian uhlans in nineteen fourteen. My father lost a leg in nineteen forty. Why should Belgium become a battlefield again because Russia and America go to war? Fight your war on your own territory and let us live in peace!"

"But that's exactly what you said you were afraid of, a minute ago," Green points out.

"I don't see . . ."

"You said you were afraid that a successful ABM shield would allow America to withdraw from Europe and leave you unprotected from the Russians."

"Yes, but . . ."

Grinning, Green says, "There's a flaw in that argument. If these satellites can defend the States against missile attack, they can also defend Europe. And Japan. And everywhere else, for that matter. The satellites are *global*, by their very nature. Not national."

"But they are controlled by the nation who places them in orbit," Mattingly counters. "If you Yanks spend the money to put them up, I presume you will be the ones who decide how and when they will be used."

"If the Russians allow you to get them working," Gunnerson mutters.

"But if," Green continues, "no, let's say *when* we get them working, up there in orbit, we'll use them to protect our vital interests. Western Europe is part of that. Western Europe has been a part of America's vital interests ever since we were colonies of England. There hasn't been a war in Europe that we haven't been involved in since Queen Anne's War, back in the early seventeen hundreds."

"The War of the Spanish Succession," Mattingly corrects. "Actually, the colonies were involved in the War of the League of Augsburg, in the sixteen eighties, but only peripherally."

"The point is that NATO is nothing new," Green insists. "America and Western Europe have been bound together for three hundred years, and we're not going to sever those ties now. The Atlantic Ocean doesn't separate us; it *connects* us!"

For a moment the little group is silent. Then Gunnerson exhales a blue cloud of pungent smoke and says, gesturing with his pipe, "I'd like to get back to something more important than political alliances: the question of life and death, the matter of nuclear weapons and this arms race we're cursed with. It seems to me that every time one of the superpowers comes up with a new weapon, the other side works like hell, not merely to make a weapon equal to the first side's, but to make a weapon that's better. Which forces the first side to go back to the laboratory and come up with something still more frightful . . . and so on until we finally blow ourselves to kingdom come."

"Or spend ourselves bankrupt," Mattingly adds.

"The Americans built the bomb, so the Russians had

to catch up," Gunnerson goes on. "Then the Russians built missiles, and the Americans had to catch up. Then they started putting tactical missiles in Europe, both of them, East and West. Then they started to MIRV the missiles. Now the Americans are building the MX and cruise missiles, which just forces the Russians to do likewise — only they'll build something bigger than the MX and something more dangerous than the cruise missile, I'm sure."

"Okay," says Green. "You're asking why we should extend the arms race into space, is that it?"

"That is precisely it," Gunnerson agrees. "It's bad enough to keep escalating the arms race here on earth. Why extend it into space?"

Before Green can reply, Colonel Stanley says, "Three reasons, at least. One: the arms race already involves space. Two: there's no reason to expect that the Russians won't place ABM weapons in orbit. Three: this is a different kind of arms race, a defensive arms race. It just might mark a turning point not only in the Cold War, but in all of human history."

Mattingly fixes the colonel with a glowering stare. "Can you elaborate on those points?"

"Yes, sir, I can." Stanley takes a breath, then plunges ahead.

"The arms race moved into space when the first missiles were flown. That was back in nineteen forty-four, I believe. The Germans started firing V-Two rockets at London and, later on, Antwerp. There was no way to defend against those missiles, except to destroy or capture their launching sites. Airplanes couldn't shoot them down. They traveled so fast, you couldn't even see them. A city block would explode; that was all the warning you got that a V-Two was on its way."

Nodding grimly, Mattingly murmurs, "I lost my sister and her two sons to Hitler's vengeance weapon."

19, St. Mark's Crescent, Regent's Park

Stanley stares at the old man for a moment. What is history to an American colonel is painful memory to this British lord. When he resumes, Stanley speaks in a lower, more subdued voice. "The ICBMs and other rocket-boosted missiles of today spend most of their flight in space, above the atmosphere. More than that, the Russians have been testing space weapons for twenty years. They have the capability of putting nuclear weapons in orbit any time they want to."

"But that's prohibited by the Outer Space Treaty of nineteen sixty-seven," Arlon protests.

"Sure it is. The Russians signed the treaty, all right. But at the same time they tested their orbital bombardment system. They didn't allow the booster to make one full orbit around the earth, so they stayed within the letter of the treaty. We call the system FOBS, meaning fractional orbit bombardment system. They've tested it several times. They have the hardware to place nuclear bombs in orbit and de-orbit them whenever they want to, wherever they want to."

"What good is that?" Gunnerson asks. "They'd be nothing more than sitting ducks up there in orbit."

"Would they? Suppose a few of the Kosmos satellites now in orbit, satellites that the Russians say are up there for scientific or observation purposes, suppose a few of them are actually hydrogen bomb warheads? The Russians could de-orbit them over the U.S. mainland and have them blow out their targets within a few minutes of their re-entry. Take out Washington, for example, while the President's giving his State of the Union speech. That would decapitate the whole American government at one blow."

The others sit in silence, digesting the possibility.

"More than that, though," Stanley goes on. "The Russians have also tested for many years an antisatellite weapon. It's simple and kind of awkward, but it works."

"Is that the system whereby they boost a satellite into space, maneuver its orbit so that it pulls alongside the target satellite, and then blow it up on signal from the ground command?" Mattingly asks.

"Or it can be detonated by its own internal command equipment," Stanley says. "The explosion shreds the target satellite."

"We're testing a much better ASAT system," Green interjects. "It's launched from an F-Fifteen fighter and goes straight to the target satellite. Much more flexible. Smarter."

"Andropov offered to dismantle the Soviet ASATs," Arlon says softly. "The offer still stands, I believe."

"Sure," Green says, "and we're supposed to stop production of our far better system."

Gunnerson shifts uncomfortably on his chair. "That's just what I'm worried about: why can't you agree to ban all weapons from space? Didn't Brezhnev propose such a ban a year or so before he died?"

"He certainly did," Green answers. "Right after the first flight of the space shuttle. The Russians consider the shuttle to be a weapon system."

Gunnerson starts to reply, but clamps his teeth on his pipe instead.

"Negotiating with the Russians isn't easy, at best," Green says. "Most of the time, it's impossible. A ban on all weapons in space, for example, would still allow the Soviets to keep their ASAT weapon. It isn't based in space; it sits on the ground until they launch it."

"And yours sits under the wing of a jet fighter plane," Gunnerson counters. "Surely you can work out a treaty that would prevent both sides from using those weapons — if you wanted to."

"Perhaps we could," Stanley says. "But what do we do in the meantime?"

15

The Soldier's Dilemma

GUNNERSON and the others stare at Colonel Stanley. The American slowly gets to his feet, walks back to the bar, and fills his tumbler from the bottle of Chablis in the ice bucket. Walking back toward the group, he says: "The politicians and the diplomats might be able to hammer out an agreement that prohibits all weapons of every kind from being used in space. Fine. I think maybe it would be a mistake, but I'm willing to live with such a treaty, if it could be enforced. I doubt that it could be, or that the Russians would allow it to be. But I'd be willing to abide by it, if and when the President signed the treaty and the Senate ratified it."

Sinking back into the chair between Green and the fireplace, the colonel continues: "But what do we do in the meantime? Sit on our hands? Hope that the Russians do the same? I said a few minutes ago that there's no certainty that the Russians aren't pursuing development of orbital ABM weapons. Suppose we enter into negotiations with them to draw up a treaty prohibiting the use of all weapons in space? Should we stop development of such weapons while the negotiators are talking? They could spend years at it, and during those years the Russians could be developing the weapons. When they're good and ready, they could break off the negotiations and announce that they've got the weapons in orbit. They own the sky."

"That's not likely . . ."

"I agree," says Stanley. "But it's possible. The point is, there's no guarantee that the Russians aren't testing such weapons already. And it would certainly be ridiculous for us to refrain from pushing ahead with orbital ABM weapons until and unless we can satisfy ourselves that they aren't."

"But that's just it!" Gunnerson is almost shouting. "You make a bogyman out of the Russians so that you can build the weapons you want. And they make a bogyman out of you so that they can build the weapons they want! And the rest of the world is caught up in your whirlwind, spinning on and on and up and up until we're all blown to hell by those weapons! You already have enough to kill everyone in the world ten times over. Why do you need more?"

Stanley's face pales as he faces the angry Dane. He glances at the others, then replies softly: "This would be easier to say if I were wearing my uniform, I suppose. But I'll try to say it anyway. I'll try to tell you what your own military men would want you to know. I'll try to represent all the military men in all the nations of NATO. I hope I can get through to you without an overdose of emotion, without touching any raw nerves that will make you stop listening. I think it's important that you understand what an impossible position you — all of you — have placed your military people in."

Stanley takes a gulp of the wine, then resumes, "I see a lot in the press and on television about 'the military mind.' I see movies where anybody wearing a military uniform is either a moron or a kill-crazy inhuman monster."

"That's just fiction," Green says. "Don't let it . . ."

"It's what the public sees," Stanley insists. "It's what people think of when they see that phrase 'the military mind.' Or when they hear somebody talking about 'the Pentagon.' It's what *you* think of, Dr. Gunnerson, when you calculate how many megatons of explosive power we have in our nuclear stockpiles."

Gunnerson nods grimly. "I think of overkill. I think that you have enough nuclear warheads to destroy the whole earth many times over."

"You're entirely right," Stanley says. "We do. But why do we? Because we enjoy sitting around and contemplating how many times we can vaporize the average Russian baby? Because we want bigger and bigger appropriations so that we can go home and tell our wives and kids that we squeezed another billion dollars out of the Congress? We don't get to spend that money on ourselves. The average military officer has a standard of living far below yours or that of any man here."

"I'm a university professor, you know," Gunnerson says, an ironic grin touching the corners of his lips. "My NATO income is something of a windfall."

Stanley shrugs. "My NATO income is zero. But that doesn't matter. What's important is this. When I joined the Air Force, I took an oath to protect the United States of America, with my life, if need be. Most soldiers pretend that they're not troubled by such oaths, that the words are just words, that it's part of the bureaucratic bullshit that goes with being a soldier or a sailor or an airman. But when you're sitting out in the hot sun behind a machine gun and some Lebanese fanatic is lobbing mortar shells toward you, believe me, you think about what you've sworn to do. When you're flying an F-Four into a swarm of radar-controlled antiaircraft guns and surface-to-air missiles, it's strange how those words come back to you."

Mattingly shakes his head slightly, but says nothing.

"We live in a world where no nation can maintain its independence without a military force to defend itself," Stanley goes on. "That may be sad, it may even be sinful, but it's true. Dr. Gunnerson, Mr. Arlon, you've both seen your countries overrun and occupied by foreign troops, because your own military forces and those of your allies weren't strong enough to repel such invasions."

"But we're talking about nuclear overkill," Gunnerson says.

"I'm coming to that," Stanley replies. "When Harry Truman was campaigning to be re-elected President in nineteen forty-eight and somebody in a crowd he was speaking to yelled, 'Give 'em hell, Harry!' Truman answered, 'Give me time.'"

The others chuckle softly.

Stanley resumes, "If you need military forces to defend you, you have to give those forces the tools they need — the weapons they need — to defend you adequately."

"Ah, there's the rub," Mattingly rumbles. "How much is enough? When does the defense establishment become so powerful that it *rules* the country instead of defending it?"

"An important question," Stanley admits. "I would say that in many countries, such as Vietnam, Poland, Libya, lots of others, the army runs the show. But not in any of the nations represented by us, here in this room."

The others glance at each other, then leave the colonel's statement unchallenged.

"What about Russia?" Green asks.

"No," says Arlon. "The Red Army has tipped the balance of power once or twice since the Second World War, but it has not assumed power for itself."

"Not yet," Mattingly adds, in a stage whisper.

"I won't speak about your countries," Stanley says, recapturing their attention, "or even about NATO. I'll just talk about the United States, my own country, the land and the people I belong to."

He takes another sip of wine. Then, "From my point of view, from inside my Air Force blue suit, the American people are schizophrenic. They want to be the strongest nation in the world, but they don't want to have a draft or any kind of required military service. They're scared to death of the Russians, but even more worried that the Pentagon is filled with idiots who pave their offices with solid gold. All through American history, from that Queen Anne's War you mentioned a few minutes ago right up to Pearl Harbor and even Sputnik, the American people have resisted paying the bills for the strong military defense that they say they want — until we get attacked. Then they scream bloody murder that we were unprepared."

Mattingly says, "It's a common characteristic of democracies. Kipling even wrote a poem about it:

"Yes, makin' fun o' uniforms that guard you while you sleep

"Is cheaper than them uniforms, an' they're starvation cheap . . ."

Stanley grins at the old man. "Yes, 'Tommy.' I know the poem."

"It seems to me," objects Gunnerson, "that a defense budget in excess of two hundred billion dollars isn't quite starvation wages."

"Oh, I agree," Stanley says quickly. "And I think that there *is* a lot of waste and a lot of nonsense in the Pentagon. But that's part of the price of defending the nation."

"We are drifting away from the subject," Arlon points out. "You were going to tell us why the military needs ten times the firepower it would take to destroy the world."

Stanley looks hard at the Belgian. There is not the faintest trace of a smile on Arlon's face.

"Right," says the colonel. "The thing is, as Lord Mattingly said a little while ago, nuclear weapons are *political* weapons more than military ones. There's a good chance that if we use a nuclear weapon in battle, even one, a small tactical bomb on a battlefield that's far from either the Russian or American homeland — there's a good chance that the whole arsenal of both sides will go off within twenty-four hours or less. The world would be destroyed completely."

"Then we mustn't use them!" Gunnerson snaps. "Not even one!"

Nodding, Stanley says, "I agree. But we can't do without them. We can't get rid of our nuclear weapons unless the Russians do. And neither side trusts the other enough even to begin the process of disarmament."

"But why do you need so many?" Gunnerson returns to his original point. "Why do you have to have an MX missile? Or a neutron bomb? Why is it necessary to make the weapons constantly deadlier?"

"Because the game we're playing against the Russians isn't war," Stanley answers. "It's *deterrence*. Each side wants to make certain that no matter what the other guy can throw at him, he has enough firepower remaining to totally annihilate the other side. That's what drives the arms race. In nineteen fifty the United States had long-range bombers that could drop atomic bombs on every major Russian city. The Russians didn't. So they came up with ICBMs."

"And the hydrogen bomb," Mattingly adds.

"Right. And every time one side adds something new to its arsenal, the other side has to match it or go it one better — or lose the race."

"But you make threat assessments that always assume the worst," Arlon points out. "If the Russians show one

The Soldier's Dilemma

missile, you assume they have a hundred. For every division the Red Army has on paper, you assume they have a fully equipped, fully manned division that is ready to attack at a moment's notice."

"Do you want us to make less conservative predictions?" Stanley asks. "Do you want us to err on the side of the Russians? Don't you think we know we're being very conservative in our estimates of the enemy's strength? But that's better than being naïve about the enemy. It's better than getting caught by surprise."

"Yes, I agree, to some extent," Arlon says. "But it's also very self-serving, isn't it? You puff up the Russian threat as large as possible, then you ask for the strength to counter that threat. The Russians do the same, and the arms race escalates beyond all necessity."

Stanley makes an elaborate shrug. "Okay. Suppose you're perfectly right. The alternative is to peg our armed strength at something less than the most conservative estimate of Russian strength. I think that's a dangerous way to run our defenses. Damned dangerous. We could wake up one day and find Soviet bayonets poking at us through our bedsheets."

"So we would be forced to surrender. Would that be so bad?" Gunnerson asks. "We'd be alive, at least. The world would go on."

Before Stanley can reply, Mattingly says, "Churchill could have negotiated terms with Hitler after the fall of France in nineteen forty. Where would Denmark be now? A vassal state of Nazi Germany. Would you prefer that to where you are today?"

"I might prefer it to being killed by a nuclear war," Gunnerson says.

Green steps into the discussion. "Listen, no matter what our individual philosophies may be, our governments and their people determine what we will or won't do. I think

Colonel Stanley hit it squarely on the head when he said the people are schizzy: they don't want the Russians to take over the world, but they don't want nuclear Armageddon, either."

"Then we must walk the tightrope between the two," says Mattingly.

"Which brings us back to this matter of ABM satellites," Stanley says.

"Indeed."

"These satellites could be a way out of the trap you're so worried about, Dr. Gunnerson," says Stanley. "They could break the spiral of constantly growing, constantly more destructive offensive weapons systems."

"For a few hundred billion dollars," Gunnerson mutters.

"Aha!" Stanley points a finger at him, grinning. "There you go, worrying about the money instead of weighing the cost against the possible benefits."

The others laugh, and Gunnerson finally breaks into a lopsided smile. "All right, I admit it."

"But together with the possible benefits," Mattingly says, "we must consider the very real risks involved."

"There's one overriding risk," says Green. "If we start to put up such satellites, would the Russians attack them, and would such an attack lead to a general war?"

"A nuclear war," Arlon adds.

"The very thing the satellites are supposed to prevent," says Mattingly.

"Okay, that's risk number one," Stanley agrees. "But there's a second risk, and it's just about the same magnitude, as far as I can see. That is, suppose the Russians put up such satellites? Just the reverse of the first situation. Do we shoot them down? If we do, that may trigger the nuclear holocaust. If we don't, we might just as well surrender to the Russians, because they'll have obtained an unbeatable military edge over us."

"I don't see where that follows," says Arlon.

Stanley replies, "Okay, let's look at the possibilities. Suppose the Russians put up enough satellites so that they can defend themselves quite well from an American missile attack. They are confident that they can shoot down almost all of our Minuteman, MX, and submarine-launched missiles. So they call Washington on the hot line and give us an ultimatum: get out of Europe or we'll launch an attack that will wipe out the United States. What do we do then?"

"You still have your bombers, your B-Ones and B-Fifty-twos. And they carry cruise missiles, don't they?"

"Yes, but is that enough to keep the Russians from handing us the ultimatum? Could the bombers get through Soviet air defenses? Some of them could, I'm sure. But enough to deter the Kremlin from attacking? I don't think so."

Puffing moodily on his pipe, Gunnerson says, "I don't think either side could possibly attack the other, not if sane men are in control. With what we know today about the global environmental effects of a large nuclear strike — why, even if only one side made an attack, it would damage the planet so badly that it would be tantamount to suicide."

"You mean this nuclear winter business?" Green asks.

Gunnerson nods.

"Do you believe that stuff?"

"Most scientists do, on both sides of the Iron Curtain."

"But do the decision-makers in the Kremlin and the White House believe it?" Mattingly asks. "That is the critical question."

"Neither Washington nor Moscow has shown the slightest intention of dismantling its nuclear arsenal because of this nuclear winter," says Arlon.

"So the picture I've just painted is still a valid one,"

Stanley says. "If the Russians get a defensive shield into orbit, they'll feel safe enough from our retaliation to attack us with everything they've got."

"Therefore your response would be . . ."

"No choice at all. We either give in to the Soviet demands or have America blown off the face of the earth. Our retaliatory blow might cause some damage to the Russians, but not enough to prevent them from launching their attack in the first place. Deterrence will have failed. Even if their defenses don't work at all, even if the satellites are a total failure and we wipe the Russians off the map, deterrence will have failed. It'd be Armageddon."

"You're saying that deterrence depends on your enemy's conception of what you can do," Green says, "rather than on reality."

"Exactly. A philosopher might say that there is no reality, only our individual conceptions of what reality is. The picture you keep inside your head is, for you, the ultimate reality. If the people in the Kremlin believe that ABM satellites will protect them from an American counterstrike, they will be tempted to launch a nuclear attack on the United States, no matter whether those satellites perform well or not."

Gunnerson takes the pipe from his mouth. "So that's why you military people make such conservative judgments of the other side's strength. That's why the arms race is constantly escalated."

Stanley nods unsmilingly.

"This scenario you have just drawn for us," Arlon says, "ends with the Russians attacking America while their satellites stop the American counterstrike."

"Right. Some of our bombers might get through, with their cruise missiles. Either America is destroyed and Russia is untouched. Or both sides are wiped out. It's the end of the world."

"For you," Arlon points out. "Europe might be spared."
"Maybe," Stanley admits. "I wonder."
Gunnerson muses, "The superpowers exterminate each other in half an hour, and the rest of the world escapes unscathed."
"Unless Sagan and Ehrlich are right about the nuclear winter," Green points out.
"Actually, Sakharov came up with that idea first," Gunnerson says. "Just as he came up with the hydrogen bomb before anybody else."
"But surely you would do something about the Russian satellites long before that point is reached," Mattingly says. "You wouldn't allow them to cover the sky with ABM weapons."
"You're probably right, sir," says Stanley. "But suppose the Russians have only a few such satellites in orbit. Maybe they disguise them as new Salyuts — space stations. Let's say they have three or four of them up there, armed with very high-power lasers or particle-beam weapons."
"Not enough to defend their homeland against a full-scale American missile attack."
"No, but sufficient to destroy our space shuttles. Sufficient to knock out our communications satellites. Not just the military satellites, but the commercial ones as well. What happens to the United States when the long-distance phone service suddenly disappears? The country's day-to-day business grinds to a halt very quickly. Chaos. No planes could fly. Trucks carrying food across the country would stop rolling. Radio and television . . ."
"Surely Washington would mobilize its forces."
"How? The phones are out! The country's paralyzed."
"That's rather exaggerated, don't you think?" asks Mattingly.
Stanley breaks into a grin. "A little. Not all that much."
"I don't think the Russians could complete two or more

such satellites without the West learning about it," Arlon says, with quiet assurance. "Our intelligence services are not without their sources. It would be far too big an operation for even the Russians to keep secret."

"Then the question becomes, What would we do about it? Negotiate with them, threaten them, or use the American ASAT weapon to destroy their satellites before they're complete and ready for operation?"

Green says, "If the Russians are building these laser weapon systems on their Salyut space station, then you can't destroy it without killing Russian cosmonauts."

"That would be an extremely risky step to take," Arlon says. "Nuclear war would be a hair's breadth away."

"Then we must negotiate with them. Make a treaty that prevents either side from placing such weapons in orbit," says Arlon.

Stanley smiles at him. "How about the opposite approach? Welcome the idea of this new arms race. It's a *defensive* arms race, after all. Tell the Russians, tell the whole world, exactly what we're doing. And tell the world whatever we learn about what the Russians are doing. Put as many satellites as we can into orbit . . ."

"But you just said that would trigger the war!" Gunnerson snaps.

"No," Stanley says firmly. "I said that if one side is doing this while the other is not, it could trigger nuclear war. It's the imbalance that creates the hazard."

"The imbalance?"

Green breaks into a wide grin. "I get it. Like the situation in the fifties, when the Russians were flight testing ICBMs and we had nothing bigger than a glorified V-Two rocket. We worked like hell and caught up with them . . ."

"You went 'way ahead of them," Gunnerson says.

"But not so far ahead that we felt totally superior," Stanley says. "Not so far ahead that we thought we could attack the Soviet Union without getting hurt badly ourselves."

"The balance of terror," Mattingly mutters.

"Right. Suppose we could turn it into a balance of safety, a balance in which both the United States and the Soviet Union feel safe from nuclear attack?"

"It would destroy NATO," Arlon insists. "It would allow the United States to retreat entirely from Western Europe. It would give the Warsaw Pact forces a free hand to pressure us. Their conventional forces outnumber us three to one, at least."

With a winking glance at Green, Colonel Stanley asks the Belgian, "But I thought that most of those divisions were merely paper? That the Pentagon overestimates the Russian threat?"

Arlon counters, "If the United States withdrew its forces from NATO, the Russians would fill out those paper divisions quickly enough. They would soon be prepared to march from the Elbe to the Rhine, and then to the English Channel."

"But the United States would still be committed to the defense of Western Europe," Green says.

"Perhaps so," argues Arlon. "But the key to that defense is your threat to use nuclear bombs against the Russian homeland. If these ABM satellites shield the Russians against such attack, then there is no deterrent against their invasion of Western Europe."

"That's why you see the ABM satellites as a threat to NATO," Green says.

"Yes, but . . ." Arlon hesitates.

"But what?"

"But there are other considerations," Mattingly replies. "Powerful considerations."

"That is true," Arlon admits.

Gunnerson shakes his head wearily. "If the Russians really are building such weapons in space, I suppose you have no choice but to stay abreast of them."

"Colonel Stanley is correct," Arlon says reluctantly. "It is not the absolute level of forces that is dangerous so much as an imbalance between the forces. If one side builds ABM satellites, the other side must, also. Regrettable. Expensive. But inescapable."

"I still believe we can negotiate a workable treaty," Gunnerson insists. "Every effort should be made."

"Yes," Mattingly agrees. "But, in Churchill's phrase, 'We arm to parley.' If we try to negotiate a treaty banning all weapons from space while the Russians are building their weapon and we are not, we will get nowhere."

"If only we knew for certain what the Russians are doing!"

"I suppose there are men in the Kremlin this very night asking themselves what the West is doing."

"If only we could trust each other."

Mattingly huffs. "Not in this century. Not in all of human history."

"Maybe someday," Stanley muses. "Maybe in the future we'll be able to trust one another."

"After the millennium," says Mattingly.

Stanley grins. "That's barely fifteen years away."

Green gets up from his chair and busies himself filling everyone's glasses. The others are silent, each of them digesting the information and ideas they have shared this evening.

Finally Green resumes his chair. "Uh . . . there's one factor that we haven't discussed, and I'd really like to have your feelings about it, if you don't mind."

"What is that?" Mattingly asks.

"Money," says Green. "If the United States decides to go ahead with development of this ABM system, would the governments of the NATO alliance be willing to join the program and share the costs?"

A deep silence fills the room for many moments.

"It always comes down to this, doesn't it?" Gunnerson says at last. "At the bottom of every move we make, there's always the question of money."

"We're talking about total system costs of several hundred billion dollars," Green says. Then he adds, "Spread out over more than ten years, of course."

"None of us can commit our governments to a course of action," Arlon says.

"I know. I just want a feeling. What do you think your governments would do?"

Mattingly asks, "What is the quid pro quo? What do we get in return for helping to fund the project?"

"Protection, for one thing," says Green. "Any nation that helps to build the satellite system will be protected under its umbrella."

"Do you mean that if a nation doesn't help pay for it, you won't defend them?" Gunnerson asks.

"I didn't say that."

"But you implied it. That's blackmail!"

"Look," Green says, raising his hands in a don't-blame-me gesture, "I'm just trying to get some information. Lord Mattingly asked a question, I gave an answer. That doesn't represent official U.S. policy, for God's sake."

Mattingly interrupts the growing argument. "I assume that American treaty obligations will apply to the satellite umbrella as well as all other American military forces, and the United States has reaffirmed its willingness to defend Western Europe many times over."

Green casts him a relieved and grateful glance.

"What I had in mind, however, when I asked my quid pro quo question, was something else. If Her Majesty's Government puts in an odd billion or so here and there on your satellite program, what measure of control will we have over the command of the satellites once they are in place?" Mattingly hikes a stubby thumb toward the ceiling.

"Control?" Green asks.

"Yes. If we help to pay for them, and I presume we will also help with some of the scientific research and development, will we share in their control and use?"

"You mean a joint NATO command for the satellite system?" Stanley asks, frowning slightly.

"Yes, of course. You can't ask us to help foot the bill without expecting us to share in the use of the final product."

"That's a decision that can be made only at the very highest levels of government," Stanley says.

"I should think so," Mattingly agrees. "But it is a decision that must be made before the first step toward international cooperation on the project can be taken."

The others all nod their heads, even Green. Only Colonel Stanley seems to be worried about sharing control of the satellites with the governments of the NATO nations.

Mattingly catches the expression on the American's face. "Come, come, Colonel Stanley. You can't very well expect your friends to help you build this shield without their insisting on the right to help decide when and how it should be used."

"I know," Stanley admits. "But still — how can we possibly have multinational control of a system that's got to respond to the enemy within seconds?"

Gunnerson, who had been trying to relight his pipe, takes it from between his teeth and says, "That's easy. Since the system will be directed by computers, just have everything preprogrammed by technicians from each of the NATO nations."

"But who decides when to start shooting?" Green asks.

"The system decides," Gunnerson says, a hint of acid etching his voice. "Once we put such weapons into space, they will have to be able to react far faster than human beings can."

"Surely human beings will still retain control of the machines," Mattingly insists. "The system can be built so that nothing can start until a human command from the ground puts the machinery on an alert status."

"Yes, that's the way it would have to be done," Stanley agrees.

"Very well then. Human beings remain in charge."

"Until the shooting starts," Arlon says.

The others fall silent.

PART IV

The Path to Peace

Life persists in the middle of destruction. Therefore there must be a higher law than that of destruction. Only under that law would well-ordered society be intelligible and life worth living.
— Mohandas K. Gandhi

16

Scenario Three: A Good Strong Roof

31 DECEMBER 1999: Perhaps they thought we would be off our guard on New Year's Eve, or drunk, or working with just a skeleton crew. Maybe one of *them* was drunk, the one who gave the order to launch the missiles. Drunk with power, most likely. Or maybe he was an ideologue who suddenly decided to end the old millennium with a blaze of glory.

Whatever, why ever, they launched their missiles. More than a thousand of them, each of them carrying from three to ten warheads. Something like six thousand hydrogen bombs came hurtling toward us from over the North Pole and from the seas on all sides of us.

We were not drunk. We were not caught napping. Thank God for that! And thank God, too, for the lasers and other defenses we had up in orbit. Without them we would have been unable to do anything to save ourselves. We would have had to let them blow us off the face of the earth, with no response except to launch all *our* missiles and blow them off the face of the earth, too. I don't think the human race could have survived that.

But we had the lasers in orbit, and a lot more, besides. For years we had worked to build a good strong roof over our people, and when the time came to test it, that roof held firm.

Our "roof" consisted of more than 250 satellites, several hundred miles up, in orbits that allowed their sensors to see just about every square inch of land and sea in the world. Most of those satellites were armed with 25-megawatt chemical lasers; huge things, they weighed a hundred tons apiece. The laser works on energy generated from the chemical reaction of highly corrosive fluorine and deuterium, the heavy isotope of hydrogen. Mix those two together inside a properly built laser cavity and out come powerful pulses of infrared energy, strong enough to punch a hole through the skin of a ballistic missile.

The laser beam is pointed by a shiny copper-plated mirror, about twenty meters across. To move the beam, swivel the mirror. The rest of the satellite remains stationary as it rides along its orbit. It's quite a technical feat to send a 25-megawatt beam across a thousand miles of empty space to hit a speeding ballistic missile. But our engineers and technicians had done their work well; even though only a dozen satellites were in the right position to fire at the missiles as they boosted up above the atmosphere, those dozen lasers knocked out more than six hundred missiles in the first six minutes of the engagement. And as the surviving boosters lit up their second-stage rocket engines, they came within range of more of our satellites, which destroyed all but seventy-six of them.

Our surveillance satellites, up in much higher orbits, studied the situation and reported back to us in our command headquarters, deep under the mountains. We tracked the remaining seventy-six missiles as they coasted toward us. Even though the satellite sensors no longer had the hot, bright plumes of rocket exhausts to lock onto,

Scenario Three: A Good Strong Roof

their infrared "eyes" showed us clearly the upper-stage buses that carried their deadly cargoes of hydrogen bomb warheads. Each warhead was a hardened re-entry vehicle, built to fly through the blazing heat generated when it dove back into the atmosphere at a speed of Mach 20 or more. Not even our most powerful laser could hope to burn its way through such a warhead's heat shield.

But our roof did not have any holes in it. Scattered among our laser-armed satellites are a few satellites that bear powerful particle-beam projectors, and others that carry small hypervelocity missiles. Quite automatically, these satellites located and locked onto the elusive seventy-six. Their on-board computers sent a query to our underground computers with the speed of light: Should we attack these targets? Our command computers are programmed to respond positively unless a human decision-maker intervenes. None of us lifted a finger. We watched our screens as the backup satellites went to work on the warheads.

We could see nothing spectacular. The warheads did not blow up, as they do in science fiction movies. But when they re-entered the atmosphere, they were no longer dangerous hydrogen bombs. Those which had been hit by the missiles were nothing but bits and pieces of junk; they burned up in the atmosphere, producing a spectacular but harmless shower of "shooting stars." Those which had been destroyed by the particle beams came right down to the ground; one of them hit a factory building like a stone and flattened it. But the bombs inside them were dead, destroyed by the incredible energies of those streams of subatomic particles.

The enemy's attack had been almost completely stopped. Three live warheads got through; they destroyed a few of our missile silos, far out in the wilderness, away from any towns or people. Our roof had leaked a little, but not dangerously. We all went slightly crazy, yelling our heads off and clapping each other on the back; after a half-hour of nerve-racking tension, it was the least we could do to let off steam.

As the dawn rose on a new year, a new century, and a new millennium, we slowly realized that the worst thing any of us had ever expected to see had happened — and we had all survived it. Even the enemy had survived it!

Now it was up to the politicians. The enemy had fired his missiles and done us no harm, except for that one demolished factory and a few missile silos. The people were delirious with exhilaration. The Cold War was over, at last, and World War III was finished before it had even begun. They had used their missiles. We still had almost all of ours sitting safely in their silos, and still more aboard our nuclear submarines. They had nothing left — and no space defenses to prevent us from obliterating them any time we felt like it.

The streets of Moscow, Leningrad, every city in the Soviet Union, filled with wildly happy people. When our delegation drove from the Kremlin to the airport, on their way to dictate peace terms in Washington, the people turned it into an impromptu parade, lining the streets and roadsides every inch of the way, strewing their path with flowers, despite the bitter January cold.

In America, of course, there were no celebrations. But I think even the Americans should be grateful that we built a good strong roof to protect ourselves, and that it was not necessary to destroy the world in a nuclear holocaust.

17

Where We Stand

IN A WORLD where "peacekeeping" forces are shot at, bombed, or ignored by armies and politicians alike, what chance is there to create a genuine international organization that has the power to prevent war?

Attempts to substitute law for violence are not new. In the year A.D. 989 the Roman Catholic Church decided to take steps to prevent warfare in Europe. Church councils, held at Charroux, Narbonne, and Puy in France, proclaimed the Peace of God and attempted to set up rules that would protect peasants, townspeople, and the churches from the constant fighting that racked the times. It decreed that clerics, monks, and nuns were inviolate; those who injured them would be excommunicated. Later, shepherds, children, merchants, and travelers were added to the list of those who must not be harmed. Certain places were named as asylums, and no man was allowed to carry arms into them.

The Peace of God failed. Violence and private wars among the nobility continued unabated.

In 1027, the Church proclaimed the Truce of God, which

established a "demilitarized zone" of time: there was to be no fighting from Saturday noon until Monday morning. By 1042, this period had been extended from Wednesday evening until Monday and included Lent and Advent. During these portions of the week and holy seasons, said the Church, "no man or woman shall assault, wound, or slay another, or attack, seize, or destroy a castle, burg, or villa by craft or violence." Bishops in the various dioceses tried to enforce the Truce of God with both temporal and spiritual punishments, but the truce failed. Indeed, many historians have considered that when Pope Urban II preached the First Crusade against the Saracens, in 1095, one of his main motivations was to set the bloody-minded warriors to fighting a common enemy instead of their fellow Christians.

The Church's peace movement of the tenth and eleventh centuries worked only when the fighting energies of Europe were turned against an exterior foe, Islam, the hated infidels in the Middle East. And even then the success was hardly complete.

In 1139 the Second Lateran Council tried to outlaw a particular weapon — the crossbow — instead of outlawing war in general. The crossbow was considered a heinous weapon. With a crossbow, an ordinary foot soldier or peasant could kill a mounted, armored knight. Grossly unfair, as far as the nobility was concerned, and it was the nobility who made up the ranks of knighthood — and priesthood. But the Church's stricture was ignored, and the "inhumane" crossbow continued to puncture the armor of Europe's chivalry until gunpowder was introduced to the West and blew away the medieval society forever.

Peace movements have had a long, and mostly sad and frustrating, history. As recently as 1928 the Kellogg-Briand Pact of Paris, formally titled the Treaty for the Renunciation of War, which was signed by sixty-three nations, renounced war as an instrument of national policy. The pact did permit the signatory nations the right of self-defense, however. Eleven years later World War II began.

Eugene V. Rostow, Sterling professor of law at Yale University and former director of the Arms Control and Disarmament Agency, observed, in an article in the *New Republic* (20 February 1984), that

> the Western publics expect too much of arms control agreements. They have forgotten the wise warning of Salvador de Madariaga, who lived through the futile efforts of the League of Nations to achieve disarmament before World War II. Disarmament can be achieved, de Madariaga said, only in a generally peaceful world. "Otherwise, disarmament conferences invariably result in agreements to have more arms."

The difference between those earlier attempts to legislate an end to war, or at least a limitation on armaments, and today's situation is twofold:

1. In a world of growing nuclear arsenals and growing instability, a major war could annihilate the entire human race. Therefore, the need to find a way to prevent war has become fundamental to the survival of humankind.
2. The new technologies of satellites, computers, global communications, energy-beam weapons, and smart weapons offer the technical means for enforcing world peace.

The technical means for warfare suppression are at our fingertips. But the political means for using these tools wisely are not. This is no surprise. Scientific and technical advances have always led social and political thinking. Consider the matter of the crossbow. There was a technical invention that brought the armored, mounted knight to the level of the peasant foot soldier. It was the first step toward a more egalitarian society; a bloody step, indeed, but a step that led after many centuries toward capitalism and modern democracies.

Jet-powered airplanes and modern electronics have changed the world more than even the hardiest of tourists can imagine. They are transforming national economies into an interactive, interdependent world economy. Robots in Japan build automobiles that are sold in the United States, and American automobile workers are laid off because they cannot compete successfully against those robots. American computers are selling throughout the world — except for Soviet bloc countries, where their sale is restricted by political policy. Nonetheless, the Russians use every means at their disposal, from "straw men" buyers to outright espionage, to get their hands on the latest computer technology. We take it for granted that we can see live television from anywhere in the world, and buy goods ranging from chocolates to shoes to TV sets to furniture from all around the globe.

Our politics lag behind these technological and commercial advances. Legally, the world is still an aggregation of separate sovereign nations, each within its own borders, each claiming territorial rights that start at the center of the earth and extend out into space, each clinging to the right to defend itself from other nations and to make war whenever and wherever it wants.

There is the United Nations, but it is helpless in the face of international power politics. The U.N. has been unable to forestall wars among small nations or large: Iraq versus Iran, India versus Pakistan, Israel versus all its neighbors, Argentina versus Britain, the United States versus North Vietnam, various upheavals in Africa. The U.N. has neither prevented them nor stopped them once they started. This is because powerlessness was built into the U.N. at its very outset; the big powers, including the United States, were unwilling to give the U.N. power that would threaten the war-making and/or self-defense prerogatives of the sovereign nations.

There are associations of nations, such as the North Atlantic Treaty Alliance (NATO), the European Economic Community (often called the Common Market), the Warsaw Pact alliance, the Organization of American States (OAS), the Organization of Petroleum-Exporting Countries (OPEC), the Organization of African Unity (OAU), and others. Although the media tend to treat NATO primarily as a military alliance, it has important scientific, industrial, and commercial aspects, as well. Yet these are all voluntary associations; even an Eastern European nation has the right to withdraw from the Warsaw Pact — on paper, at least.

If the trend of history continues, as technology makes the world smaller and ties the nations of the globe together into an ever-tighter interdependency, national governments will gradually give way to multinational arrangements. We can see this happening today. In fact, some observers of the modern scene have pointed out that multinational corporations, many of which have annual operating budgets that are bigger than the gross national product of most nations, are already exerting a form of international governance in the Western and third worlds. Multinational corporations and banks also do a surprising amount of business with Soviet bloc nations. Perhaps someday the world will be ruled de facto by such corporations.*

If this sounds strange or repugnant, remember that the citizens of glorious ancient Athens never conceived of the nation of Greece, nor did the barons of Burgundy or Cornwall particularly like the idea of giving up their power to the kings of France and England. Patrick Henry gravely distrusted the idea of a strong central government for the

* I wrote a novel, *Colony*, about such a future.

United States, as did many of the patriots who fought the American Revolutionary War. If we survive into the twenty-first century, we may see a world in which national governments still exist, still have the de jure power in the world, but play a secondary role to the border-crossing, apolitical corporations.

Even if that happens, though, the last thing that the nations would give up would be their right to self-defense; that is, their right to make war.

While virtually all sane people agree that nuclear weaponry has brought humankind to the brink of extinction, and that some means better than MAD must be found to assure that nuclear war will not break out, hardly anyone has a clear vision of how to achieve a world in which the nuclear weapons have been dismantled and war is no longer a possible option for settling international disputes.

We know where we want to be: in a world that is free from the fear of war. The question is, How do we get there from here?

The shift in strategic thinking from MAD to Assured Survival may offer the means.

In the September 1982 issue of *Omni* magazine, I was happy to publish Gregory Benford's suggestion that what we need is a new arms race — a *defensive* arms race.

> Unlike offensive systems [Benford wrote], the addition of a new [defensive] system makes the situation more stable. Battle stations in orbit cannot . . . be used offensively against any nation. At worst, they can destroy other satellites, but we already live with antisatellite weapons, so that doesn't alter the picture.
>
> It is time to quit deploring all arms races. We should try to distinguish between suicidal offensive systems and shrewd defensive ones. True . . . mankind would be wiser to negotiate a way out. But . . . humankind will continue to arm, if history is any guide. Let's do it intelligently.

Benford, with his physicist's training and science fiction writer's vision, had presented the kernel of a new idea.

Stewart Brand, editor and publisher of *The Whole Earth Catalogue* and *Co-Evolution Quarterly*, had nibbled at the edges of the idea in a tongue-in-cheek article he wrote for the British magazine *New Scientist* in 1981. Titled "War in Space: Good!" Brand's piece started: "Remember when space was supposed to be like Antarctica, with its exemplary treaty declaring that this continent is for science, no fighting allowed?" He then said that "we peaceniks . . . are trying . . . to declare space a no-fire zone, sacred, taboo, for science only (and maybe a little commerce), a better, cleaner place, free of old bad habits."

But, Brand pointed out, such thinking may be

dead wrong. . . . Space is *perfect* for war. And Earth is increasingly lousy for it. Earth is crowded with people and other organisms. . . . And proper weaponry, the inexorable product of technology peristalsis, simply is not well suited to a planet's surface, especially one with weather and running water. . . .

I had this vision the other day. . . . War becomes increasingly a space activity, conducted ever farther from stodgy old Earth, using materials and energy direct from asteroids and the Sun. Then one day in an ultimate shoot-out both sides send out immense fleets of perfectly matched killer ships. Off they go in a ferocious soaring battle, blasting and lasering and poisoning and spying and inventing and reconstructing and re-attacking — a furious blazing war. *And it just keeps going* — off into interstellar dimness, gone. None of the ships ever returns. And so humanity wakes from its adolescent possession, the demon is exorcized, sent fizzing and blaspheming off into the reaches of space.

The devil theory of war may not make much sense, but in his own way Brand has made something of a point: weapons in space *can* help to eliminate war from earth, even if not in the whimsical way Brand wrote about.

In fact, efforts are already under way to utilize space in peacekeeping efforts. These efforts actually began with the work of a single husband-and-wife team, Howard and Harriet Kurtz, who, for more than thirty years, have urged a "global information cooperative" that would share with every nation in the world all the commercial, environmental, and military information recorded by all the surveillance satellites in space. Harriet Kurtz was an ordained minister of the United Church of Christ; Howard, a former U.S. Air Force lieutenant colonel and an engineer. They began their crusade for international cooperation in the mid-1940s, when Howard, then an employee of American Airlines, tried to establish an air service route between New York and Moscow. Cold War politics prevented the air service from getting under way — and convinced the Kurtzes that international cooperation had to be based on the free exchange of information.

Year after year the Kurtzes held round-table meetings to discuss various ways of utilizing existing technology to build international cooperation. They formed War Control Planners, Inc., and suggested that NASA and the Department of Agriculture form a Central Agricultural Intelligence Agency to disseminate, worldwide, data on crops and weather conditions. Their efforts seemed to go nowhere at all. But among the thousands of men and women with whom they spoke or corresponded, some listened and understood.

In May and June 1978 the United Nations General Assembly held its first Special Session on Disarmament. As usual, there was more rhetoric than progress. But the French delegation proposed the establishment of an International Satellite Monitoring Agency (ISMA), which would place satellites in orbit for the purpose of monitoring arms control agreements and observing the scenes of international crises so that accurate and up-to-the-minute information could be presented to the U.N. and the world. Since many U.N. resolutions have stressed the need for effective verification of agreements between nations, especially in the area of arms limitations, a U.N.–controlled network of surveillance satellites seemed a useful idea. The seed that the Kurtzes had planted had finally begun to sprout.

France has been one of the most active nations in Western Europe in the field of space technology. In addition to its membership in the eleven-nation European Space Agency (ESA), France has been a major contributor to the development of ESA's Ariane rocket booster and the creation of the semiprivate commercial endeavor, Arianespace, which sells launching services to communications companies and others who pay to place satellites in orbit. France has undertaken several joint space ventures with the USSR, and in 1982 Jean-Loup Chrétien became the first French *spationaute,* spending several days in orbit with three Russian cosmonauts aboard the Salyut 7 space station. The French have also developed the finest civilian satellite surveillance system in the world: SPOT (Système Probatoire d'Observation de la Terre), which offers better resolution than the American Landsat system. SPOT technology is the basis for the French ISMA proposal.

At its thirty-third regular session, the General Assembly "requested" the Secretary-General to commission a study of the technical, legal, and financial implications of an ISMA that, in the language of the diplomats, could make an "important contribution . . . to the solution of monitoring problems, taking into account, in particular, the need to provide for international measures which are nondiscriminatory and do not constitute interference in the internal affairs of States." In other words, surveillance satellites could allow the U.N. at least to observe what individual nations are doing to comply with U.N. resolutions, with bilateral agreements, and during periods of international crisis. The satellites would not require foreign inspectors on a nation's soil, and since the superpowers had long since agreed not to oppose "national technical means" of observation (meaning satellites), an ISMA might even get past the inevitable squabbling between East and West.

Secretary-General Kurt Waldheim appointed a Group of Experts, which was chaired by France's Hubert G. Bortzmeyer, of the Centre National d'Etudes Spatiales (National Center for Space Studies). The group met six times during 1979 and 1980 and worked closely with the U.N. Center for Disarmament, the U.N. Outer Space Affairs Division, the International Telecommunications Union, the World Meteorological Organization, the International Atomic Energy Agency, the Stockholm International Peace Research Institute, and the space experts of many nations. By the end of 1981 the group had unanimously adopted its report, which was presented to the General Assembly later that year.

The report concluded that an ISMA could make a "valuable contribution . . . to the verification of certain parts or types of arms control and disarmament agreements." Satellites could also play a "positive role . . . in preventing or settling crises in various parts of the world and thus contributing to confidence-building among nations." The group saw no technical problems with the idea of orbiting a network of surveillance satellites, and no legal impediments to establishing an international monitoring system.

In December 1982 the General Assembly voted to adopt the group's report; 126 nations voted affirmatively, 9 voted no, and 11 abstained. Among the negative votes was the Soviet Union's. Among the abstentions was the United States'.

"The U.S. has nothing to gain from ISMA," says Jerry Grey, author of *Beachheads in Space* and publisher of *Aerospace America*, the monthly magazine of the American Institute of Aeronautics and Astronautics (AIAA). The American government's position, he explains, is that "we already have the capability [of surveillance satellites]; why should we pay to allow other nations to acquire the capability?"

But Grey does not fully agree with the official United States position. Although he realizes that the United States would have little to gain from ISMA, and that the immediate release of military information gleaned by surveillance satellites might actually be *de*stabilizing in a crisis situation, he firmly backs the AIAA's recommendation that the United States openly publish its intelligence photographs of the USSR, rather than keep them secret. Grey and his fellow aerospace scientists and engineers see this open-publication policy as a step toward making the world safer.

Arthur Clarke, long an advocate of using space to enhance peace on earth, found himself arguing in favor of the ISMA idea while on a visit in 1982 to the Soviet Institute for Space Research, in Moscow. One of the Russians remarked sarcastically that the whole scheme was merely a ploy by the French to sell their own SPOT equipment. Clarke was furious with himself for not thinking of the proper retort until several days later: "So what?"

Clarke has lived in Sri Lanka for three decades. He was instrumental in getting the Indian government to establish the first direct broadcast satellite (DBS) system in 1974, in which small, inexpensive receiving antennas and TV sets were placed in some four thousand villages throughout India while the United States placed the world's first direct broadcast satellite into a geostationary orbit over the Indian Ocean. For about a year the satellite beamed information on health care, sanitation, and farming to those villages across the vast subcontinent. The experiment was a decided success, and direct broadcast satellites are now beginning to transform the communications industry.

Clarke, of course, has a rather personal feeling about space in general and communications satellites in particular. A founding member of the British Interplanetary Society in 1934, he invented the idea of commsats in 1945, and has seen his "science fiction" idea turn into a multibillion-dollar-per-year global industry. His own prophetic novels have won him worldwide fame. Less well known, perhaps, has been his continued work in the areas of education (he is chancelor of the University of Moratuwa in Sri Lanka) and in spreading the word about the commercial, scientific, and even spiritual values of space flight.

In December 1982, Clarke summarized his views in a talk titled "War and Peace in the Space Age" at the first seminar of the Institute of Fundamental Studies, in Colombo, the capital of Sri Lanka.

"Nothing is more fundamental than the prevention of nuclear war," Clarke said. "If we fail in this, all else is irrelevant."

Commenting on the ISMA proposal before the U.N., Clarke said:

> The operational and political difficulties are obviously very great, yet they are trivial when compared with the possible advantages. The expense — one or two billion dollars — is also hardly a valid objection. It has been estimated that its reconnaissance satellites saved the United States the best part of a *trillion* dollars. A global system might be an even better investment; and who can set a cash value on the price of peace?
>
> However, the United States and the Soviet Union, anxious to preserve their joint monopoly of reconnaissance satellites, are strongly opposed to such a scheme.

Clarke dubbed the ISMA concept "Peacesat," and argued that this is an idea whose time has come.

> Whether the superpowers wish it or not, the facilities of an embryo Peacesat will soon be available to all countries. May I remind my Russian and American friends that it is wise to cooperate with the inevitable; and wiser still to *exploit* the inevitable. . . . Peacesats could develop in a noncontroversial manner out of . . . a consortium of [existing international] agencies for weather, mapping, search and rescue, resources and pollution monitoring, disaster watch, information retrieval and, of course, communications. No one denies the need for these facilities. If they were provided globally, they would inevitably do much of the work

of the Peacesat system. The only extra element required would be the evaluation and intelligence teams needed to analyse the information obtained.

He noted that the Soviet Union's COSPAS search and rescue satellite had already "saved a dozen American and Canadian lives, by detecting the faint radio signals from downed aircraft and wrecked ships (five within two months!). This is just the beginning of a system which we will soon take completely for granted, as we have done ever since the *Titanic* sent out its first distress call seventy years ago."

Admitting that ISMA or Peacesat is "not a magic solution to *all* the problems of peace," Clarke nevertheless concluded, "But at least it is worthy of serious consideration, as one way of escape from our present predicament — all of us standing in that pool of gasoline, making our Mutual Assured Destruction ever more assured."

Peacesats may present a way to help stabilize crisis situations when they occur. And, like the existing American and Soviet surveillance satellites, they can help to verify that nations are living up to their arms limitation agreements without placing foreign inspectors (spies, as far as the Russians are concerned) inside a nation's sacrosanct borders. But without the agreement and funding aid of the United States and Soviet Union, it seems doubtful that the U.N. can establish a Peacesat network. Will the other nations of the world push through a Peacesat program and fund it? If they do, it could be a positive step on the road toward lessening global tensions. It could also be a vital step toward the eventual establishment of an International Peacekeeping Force.

Richard H. Ullman, professor of international affairs at Princeton University and a visiting member of the Institute for Advanced Study, suggested in the 23 April 1983 *New York Times* that the United Nations be given authority to operate space-based ABM defenses. He admitted that most Americans see the U.N. as, at best, a waste of time which "brought us the 'Zionism is racism' resolution . . . routinely censures the United States for keeping Puerto Rico in colonial thralldom, and so often, when the chips are down, seems so impotent." But, he quickly added, "only a combined effort to develop antimissile defenses — one that joins the resources of the United States and the Soviet Union — can prevent a separate effort by one country from goading the other to develop countermeasures, thus fueling the ever more dangerous arms race."

Ullman pointed out that President Reagan himself suggested that, eventually, when the United States has the new technology working, it could offer similar ABM systems to the Soviets to prove that they no longer need to maintain their arsenal of offensive nuclear missiles. But he correctly argued that no prudent Soviet leader would be content to wait for that fortunate day; the Russians will push for their own countermeasures to American orbital ABM weapons, whether those countermeasures are ABM satellites of their own, more offensive missiles, or something else.

"But would a joint effort mean sharing the most closely held secrets of space-age research?" Ullman asked. "Of course it would. Only open collaboration could reassure either side that the other was not on the verge of a breakthrough that would suddenly enable it to render its opponent helpless."

In all fairness, even an open joint effort would probably not convince the most hawkish minds either in Moscow or Washington that the other side was telling everything it knew. But a collaborative effort on space-based defenses, if it could be undertaken, would be a crucially important step toward building the confidence and trust between the superpowers that is undeniably necessary if we are to move away from the policy of MAD. Ullman suggested that a joint Soviet-American effort could be managed by the U.N. as an impartial representative of the rest of the human race. Such a program would be far better than Benford's suggested defensive arms race. In effect, it would be an arms race in which both sides work together and share the costs.

Once the orbital weapons are ready for deployment, the U.N. would take over their operation, in Ullman's scheme. In that way, neither East nor West will have control of them, and both sides can begin to dismantle their fleets of strategic nuclear missiles.

> Finally [Ullman concluded], it should be emphasized that a United Nations missile defense, radical a change as that might be, would not eliminate the scourge of nuclear war. Aircraft, cruise missiles, and terrorists with suitcases can also deliver nuclear weapons. They may not be as rapid as ballistic missiles, but they are no less deadly. . . . Ridding the world of nuclear war implies a political fix that would be even more difficult to bring about.

Ullman's proposal illuminates another crucial part of what may one day be an International Peacekeeping Force. By internationalizing the space-based ABM defenses, and integrating them with Peacesats that monitor the nations of earth, two of the major parts of an effective IPF could be brought into existence. A third necessary part is the establishment of fleets of roboticized smart missiles, drone aircraft, tele-operated systems, that can meet and counter conventional warfare on land, at sea, and in the air.

The key to it all, of course, lies in the willingness of the nations — particularly the superpowers — to allow an International Peacekeeping Force to come into being. And there we face the most obdurate problem of all. Looming before us like the Great Wall of China stands the intransigence of almost every nation on earth toward surrendering the slightest shred of its sovereign right to wage war.

This intransigence is based on fear, of course. No nation dares to give up its arms and place its whole reliance for defense on a group of strangers.

Could the United States and the Soviet Union lead the way toward a safer world by working together on space-based defenses? All the indications are extremely negative. There is not one hint that anyone in a position of power in either nation is ready to take even the smallest first step in that direction, and plenty of evidence that both Washington and Moscow would resist tenaciously any effort to get them to share their technological "secrets."

Yet it is in that direction that the human race must head if we are to survive the age of nuclear weapons and MADness. Sooner or later, Americans and Russians will have to learn to live with one another, to work together, even to trust each other.

In *Nuclear Hostages* Bernard O'Keefe wrote, "We say we can't trust the Russians. But we trust them every day, every hour. We trust that they have the good sense and the instincts of self-preservation deeply enough ingrained . . . to resist the urge to push the button that will finish us all."

As a matter of fact, we trust the Russians somewhat more than that. And they trust us, to an extent that most people do not often realize. Many Americans and Russians have labored long years to bring their two governments to some degree of understanding and trust. Since the nuclear test moratorium of 1958–1961, the United States and the Soviet Union have ratified a hatful of bilateral and multilateral agreements that have helped to edge the two superpowers slightly away from nuclear confrontation. These include:

The Antarctic Treaty, which went into effect in 1961, was originally signed by the United States, the Soviet Union, Argentina, Australia, Belgium, Chile, France, Japan, New Zealand, Norway, South Africa, and Great Britain. Later,

Czechoslovakia, Poland, Denmark, the Netherlands, and Rumania also signed it. The treaty establishes the Antarctic continent as a demilitarized zone reserved for scientific research; the signatory nations agree not to make any claims of national sovereignty to any part of the continent. However, the treaty expires in 1991, and the government of Argentina has shown indications that it may make a claim to some Antarctic territory. (A pregnant Argentinian woman was flown to the Argentine research station in Antarctica and gave birth there, thus allowing the Argentine government to claim that since at least one of its citizens was born there, the territory "belongs" to Argentina!)

Limited Nuclear Test Ban Treaty, of 1963, signed by more than a hundred governments, prohibits testing nuclear weapons in the atmosphere, in the oceans, or in space. The two most notable holdouts — countries that have not signed the treaty — are France and China. They continue to test nuclear weapons in the atmosphere, producing measurable increases in radioactivity throughout the world.

Outer Space Treaty, 1967, signed by sixty-three nations, establishes the moon and other celestial bodies as demilitarized territories that cannot be claimed by any nation (similar to Antarctica) and prohibits placing "weapons of mass destruction" in space. "Mass destruction" weapons are understood to include nuclear, chemical, biological, and radiological weapons.

Treaty of Tlatelolco, 1968, establishes Latin America as a zone in which the signatory nations agree not to place nuclear weapons. Fourteen Latin American nations signed the treaty; Argentina, Brazil, Cuba, the Dominican Republic, and Uruguay did not.

Nuclear Nonproliferation Treaty, which went into effect in 1970, has been signed by more than sixty-two nations. Under its terms, the nuclear powers agree not to assist a nation in any way to acquire nuclear arms, and the non-nuclear powers agree not to develop or acquire nuclear weapons. Among the nations that have *not* signed this treaty are Argentina, Brazil, France, Israel, Pakistan, India, China, Cuba, South Africa, South Korea, Saudi Arabia, and several other Arab states.

Seabed Treaty, 1972, bans the placement of nuclear and other types of weapons of mass destruction on the floor of the ocean, beyond a twelve-mile territorial limit from the seacoast. Eighty-seven nations have signed this treaty.

SALT I and ABM Treaty, signed and ratified by the United States and the Soviet Union in 1972, limits the number of strategic missiles each nation may have, and limits the number of ABM missiles and the types of ABM defenses that may be deployed by each nation.

There are three important treaties that have been signed by the United States and Soviet Russia, but have not been ratified by the U.S. Senate. These are the Threshold Test Ban Treaty of 1974, which would limit underground nuclear tests for military purposes to explosive yields of no more than 150 kilotons; the Peaceful Nuclear Explosions Treaty of 1976, which would allow the use of nuclear explosives for nonmilitary purposes such as construction and mining; and, of course, SALT II, which would set limits on the number of nuclear warheads the superpowers could possess.

American and Russian delegates have engaged in negotiations on more comprehensive arms limitations in the Strategic Arms Reduction Talks (START) and the Intermediate Nuclear Forces, or Euromissiles, talks. As of this writing, the INF talks were cut off by a Soviet walkout when American Pershing II and cruise missiles began to arrive in Western Europe, in November 1983. The START negotiations are also on hold; the talks recessed in December 1983, and the Soviets refused to agree to a date for their next session.

There is one other major area of potential negotiation between the United States and the Soviet Union: a treaty banning antisatellite weapons. In 1977 the United States proposed an ASAT treaty. At that time the Soviet ASAT system had been undergoing testing for roughly a decade; two years later the U.S. Department of Defense declared the Russian killer satellite system to be operational. Talks were started, but they dragged on without much progress, and when the Red Army invaded Afghanistan in 1979 the talks were broken off. In 1981, when the American space shuttle made its first flight, the Soviet delegation to the United Nations submitted a draft treaty outlawing all weapons in space — and clearly identifying the shuttle as a weapons system. In 1982 and again the following year, Yuri Andropov, then the Soviet leader, proposed a treaty banning all antisatellite weapons. Although a treaty that would forestall the militarization of space has some support in the Congress and among citizen groups of prospace activists, the White House has remained decidedly cool to the idea, mainly because the Air Force is just beginning to test its own ASAT system, and strategists would rather have the weapon working before they enter into negotiations to control its use or give it up.

This is an old story in arms negotiations: Churchill's "We arm to parley." If the United States agreed to a ban on ASAT weapons when the Soviets had an operational system and America did not, the fear in the Pentagon would be that the Russians could destroy American satellites if and when they felt it necessary to do so, treaty or no treaty, but the United States would be left without an operational weapon and thus would have no way to counter the Soviets.

Moreover, once the idea of a space-based ABM system reached the Oval Office, interest in an ASAT treaty rapidly shrank to almost zero. Weapons in orbit that are capable of destroying ballistic missiles could easily wipe out satellites, too. There is no practical way to write a treaty forbidding attacks on satellites when missile-killing weaponry is being put in space.

The basic problem has always been a matter of balancing trust against fear. With both superpowers in possession of enough nuclear firepower to destroy the world quite literally, neither side has been able to overcome its fear of the other enough to lay down its ultimate defense: the threat of nuclear annihilation. Thus, even while negotiators were laboriously hammering out the Limited Test Ban, the Outer Space and Seabed and Nonproliferation treaties, other members of their own governments were pushing for more and bigger missiles of greater accuracy, for MIRVing ballistic missiles, for ASAT weapons, and developing the MAD doctrine.

Although the superpowers have negotiated some limitations to their strategic nuclear power, they have refrained from any steps which they see as a threat to their basic ability to destroy each other. Every effort to produce a *comprehensive* test ban, in which all nuclear weapons tests would end, has been rebuffed. Current talks about a "build-down" of missiles, in which each superpower would dismantle some of its missiles and replace them with a smaller number of newer (and deadlier) missiles might make the negotiators feel that they are making progress, but it does not end the threat of instant nuclear devastation. Even such limited agreements as SALT I have been painfully slow to achieve. By the time SALT I was signed by Nixon and Brezhnev, both nations were working feverishly on MIRV technology that was to make goals of the agreement virtually meaningless.

Meanwhile, however, the idea of international peace-keeping forces has been tried again and again, usually in the tumultuous Middle East, usually with little or no success. Time and again the United Nations has inserted peacekeeping troops — generally from small or nonaligned nations — between the belligerent forces in the Middle East. They have been political symbols, rather than an actual police power. When the warring factions wanted a truce, the peacekeeping forces were respected. When the factions wanted to resume fighting, the peacekeepers were either ignored or ordered to pack up and get out of the way. In 1983, when American, British, French, and Italian peacekeeping troops went into Beirut, they merely became targets for the sectarian militias that continued their civil war in Lebanon undaunted by the peacekeepers or a succession of cease-fire agreements. U.S. Marine peacekeepers lost nearly three hundred men in 1983, killed by snipers, mortar and artillery fire, and car bombs. In 1984, the peacekeepers were withdrawn, and the fighting in Lebanon continued.

Brian Urquhart, the U.N. General Secretary for Special Political Affairs, believes that peacekeeping operations can be made to work successfully. "I have a dream," he says, "that we shall have peacekeeping operations in which contingents from the United States and Soviet Union join those of other countries under the mandate and directives of the Security Council to keep the peace and contain conflict in crises that threaten international peace and security."

In other words, peacekeeping will become effective, in Urquhart's view, when the two superpowers can agree to make it effective. His hope is that the United States and the Soviet Union will see that it is in their best interests to cooperate in maintaining peace: a hope that is not yet justified by events.

Slowly, though, the idea of peacekeepers with enough power to enforce the peace is percolating through the political minds of the world. The former President of France, Valéry Giscard d'Estaing, proposed in April 1983 that

Western European nations could create a rapid-deployment military force that would help to maintain peace around the world. Using French, British, West German, and Italian troops under a single command structure, the force would be deployed in response to requests for help from the United Nations or regional groups such as the Organization of African Unity. Curiously, Giscard d'Estaing predicted that one of the virtues of such a multinational force would be its help in welding more closely together the West European nations.

Much of the technology and some of the political ideas for a true International Peacekeeping Force already exist. What will it take to make the IPF a reality, to rid the world of the fear of war?

18

Scenario Four: The Great Divide

As the violence-racked twentieth century drew to a close, the superpowers were clearly heading toward nuclear war. There already was a war, of sorts, going on in orbit, and it threatened to spread to the surface of earth, to the cities and homes of nearly five billion human beings.

For almost a decade, both the United States and Soviet Russia had been testing energy-beam weapons in space. Hundreds of astronauts and cosmonauts had been carried aloft in shuttle rockets and had worked for months on end in ever-growing laboratory complexes that orbited a few hundred miles above the ground. Although neither nation publicly admitted it, for the past two years both had been placing huge, hundred-ton satellites in orbit — unmanned, automated "battle stations," armed with energy weapons that could destroy ballistic missiles and other satellites with their invisible, unstoppable beams of sheer fury.

And while each side spent hundreds of billions to place these battle stations in space, it was also exerting every effort to destroy the other side's battle stations. An undeclared, silent war was being fought a few hundred miles above our heads, a war of machine against machine. The strategic planners in Moscow and Washington knew that if one side could complete its entire network of some two hundred battle stations before the other side did, it could command the situation in space — and on earth. A complete network of such satellites would protect its nation against any missile attack, from any spot on earth. It would give that nation the dominant hand, the "high ground" in this new arena of warfare, allowing it to threaten a nuclear missile strike on its enemy without fear of a devastating counterstrike.

So they played their deadly, automated game in orbit. They even kept score on their huge wall-sized viewing screens, counting how many battle station satellites *we* had in orbit, how many of them were working, how many needed to be repaired, and how many *they* had up and functioning.

There were strict rules to the game. One did not attack people. No battle station was in danger while it was being assembled in space from the components carried into orbit aboard the rocket shuttles. Once the human construction crews had left, though, the satellite was fair game for the probing, burning, slashing energy beams from an enemy battle station, which might be hundreds, even thousands of miles distant. Similarly, when human crews were sent to repair a damaged battle station, they were safe. Neither side was ready to provoke a real war; neither side dared to risk a full-scale nuclear attack, not so long as it had the chance to complete its space-based defenses. Even though the costs were backbreaking and progress was painfully slow, both the United States and the Soviet Union were placing more and more of the gigantic battle stations in orbit, together with decoys to confuse the enemy's targeters.

Scenario Four: The Great Divide

Deep inside the programming tapes of the most powerful computers in the Kremlin and the Pentagon there lay buried a magic number. No human being could calculate this number, because it depended on too many variables: the actual number of functioning battle stations in space, their exact orbits, their states of readiness (which were fully fueled; which had run low), where the ground tracks of their orbits lay in relation to known enemy missile silo complexes and submarine patrol routes, which way the wind was blowing over primary and secondary target areas (fallout is an important weapon), what day of the week it was, whether or not the Congress or the Supreme Soviet was in session, where the heads of state and the leaders of the government and military were expected to be, and many other factors. Once those constant, churning, continuing calculations produced that magic number, it would be safe to launch a full nuclear strike on the enemy. Every day, every hour, tension-racked human beings queried the computers. So far, the answer was always the same: the magic number had not yet been reached. Not yet. But every day, every hour, the computers' calculations told the drawn-faced humans that the moment was inching a little closer. Not yet, but soon. Soon. The humans stared at the computer screens, some of them obviously frightened, others disappointed.

In early April a new factor entered the computers' unending calculations: a border war had sprung up between Vietnam and China. The computers hummed and beeped, going automatically into a subroutine that contained more than a thousand separate components. But even that complexity was insufficient to deal with all the elements involved. Human logicians and programmers were summoned and began to work around the clock to update the computer programs and feed in the new information. Human strategists, military and civilian, watched, with growing apprehension, the situation in Southeast Asia through the camera eyes of surveillance satellites.

The situation was fraught with complexities. Vietnam was nominally a Soviet client, not a vassal state, as Western propaganda often described it, and certainly not a puppet. Bred by the brutal hardships of its thirty-year war for independence, Vietnamese leaders told their people that they had driven out the Japanese, the French, and finally the Americans. Then they had gone on to establish hegemony in Laos and Kampuchea. The Vietnamese Army was the largest, toughest, and most heavily armed in Southeast Asia. When the huge, dangerous dragon on its northern border attacked in 1979, Vietnam outgunned the larger but less well equipped Chinese forces and forced China to withdraw back behind the border.

Vietnam still felt the hostility of China, however, and the Soviet Union was happy to fuel Vietnam's fear and suspicion of its giant neighbor. Soviet Russia, after all, had its own border problems with China, more than two thousand miles north of Hanoi.

For its part, China watched its neighbor to the south with all the wariness and suspicion of a man watching a snarling dog. The leadership in Beijing, always an exquisite balance between differing factions, was united on this one matter: Vietnam must not be allowed to threaten Chinese interests in Southeast Asia. For centuries, whenever China had a strong central government it had obtained control over Southeast Asia, Tibet, Mongolia, and Korea. Today, only Tibet was fully under Chinese sway, and that was largely because it has almost no strategic importance or political prestige. But Mongolia, North Korea, and Vietnam were all under Soviet influence, a situation that was intolerable to the leaders in the Forbidden City. Still, the Chinese knew the virtues of patience and careful planning. They were content to wait, for they felt that time was on

Scenario Four: The Great Divide

their side. China was growing stronger, establishing ties of trade and technological commerce with the West and with the bustling little centers of industry that ringed China's coasts like a brood of noisy children clinging to their mother's skirts. In time, the rulers in Beijing reasoned, all these "new" countries — Hong Kong, Malaysia, Singapore, and particularly Taiwan — would be absorbed by China, as would their older neighbors: Korea, Mongolia, Vietnam, perhaps one day even Japan.

It was a shock, then, when Vietnam attacked at half a dozen points along the border with Kwangsi Province, capturing the city of Pingxiang in the first day's fighting. It has been said that a nation with a large and heavily armed army is like a man riding on the back of a tiger: he dare not let go, for fear of being devoured. The Vietnamese Army, which had almost created the nation — and been created by it — was a tiger from which the Hanoi government could not dismount. The army had engulfed Kampuchea and cowed Laos into submission. Its skirmishes along the border of Thailand could lead to a confrontation with the West, and no one in Hanoi wanted the hated Americans or Europeans back in his country. To the south lay Malaysia, also closely allied with the West. To the north lay Vietnam's traditional antagonist and tormentor. Long conversations between the Vietnamese ambassador in Moscow and the Soviet leaders convinced Hanoi that the Russians would not be unhappy if Vietnam embarrassed the Chinese. Perhaps those conversations were too subtle; perhaps the Vietnamese heard only what they wanted to hear. Certainly the Russians expressed as much shock as anyone when the war broke out. But they dutifully began to airlift heavy tonnages of military supplies to their Asian ally.

China, obviously unprepared for the Vietnamese onslaught, was deeply embarrassed before the world's watching eyes to have this tiny upstart nation driving a mechanized invasion into the heart of the Kwangsi region. The first reaction from Beijing was to fall back on Maoist guerrilla tactics, urging the people of Kwangsi to fight back until the main strength of the Chinese Army could be brought to bear against the enemy. But guerrilla tactics are of limited use against a tightly disciplined, mobile mechanized army. Villages were burning all along the Kwangsi border. The would-be guerrillas were little more than a tide of panicked refugees, clogging the roads leading northward, fleeing as the enemy advanced.

The bigger an army is, the longer it takes to mobilize it. The Chinese Army, concentrated mainly far to the north along the border with the USSR, needed days, perhaps weeks, to reorganize itself and reach the fighting front. The units that were on hand in the south had been shattered in the first few days of fighting as the Vietnamese overwhelmed them with superior firepower, mobility, and discipline. In Beijing, the Central Committee sat in continuous session, desperately seeking a way to stop the Vietnamese advance quickly.

It was the army's chief of staff who suggested using nuclear weapons. The Premier and his Deputy were divided over the question, as might be expected. But the Central Committee's vote was overwhelmingly in favor of the general. Since they were not prepared to use battlefield or tactical nuclear weapons on their own soil, however, the discussion quickly focused on how best to use the weapons that would bring the invading Vietnamese to their senses: hydrogen bombs sitting atop ballistic missiles.

The argument raged all through the night and past noon of the following day. When it was finally concluded, the Deputy Premier had offered his resignation (which was quickly accepted) and the general had received permission to wipe out Haiphong. No warning to the Vietnamese would be issued beforehand. They had attacked China

Scenario Four: The Great Divide

without warning; they had sown the wind and now would reap the whirlwind.

It took a full day to adjust the guidance system aboard the rocket booster in China's main missile center, deep in the rugged mountains of Xinjiang. At precisely three minutes before midnight, the rocket engines roared to life and their payload of mass murder hurtled up and into space, on its way to Haiphong. Beijing simultaneously notified its ambassadors in Moscow and Washington to warn the Russians and Americans that this one missile was the only nuclear warhead the Chinese intended to use — unless Vietnam refused to leave Chinese territory.

At 12:21 A.M., local time, Haiphong harbor exploded as the 400-kiloton hydrogen bomb aimed at the center of the city scored a near miss. Still, the shock wave from the explosion blasted down buildings as far as three miles away, and the cloud of radioactive steam raised by the explosion killed approximately seventeen thousand men, women, and children, either by outright scalding or through massive overdoses of radiation.

Making the best of the situation, Beijing announced that this had been merely a demonstration, and that Haiphong, Hanoi, and every city in Vietnam would be leveled unless the invading army returned to its own country. Moscow and Washington had already sent messages urging restraint to Beijing. Hanoi erupted with predictable fury — but the headlong advance of the Vietnamese Army halted. The soldiers received orders to dig in where they were in the best defensive positions possible and await further developments.

For the moment, the automated war in orbit between the United States and the Soviet Union was halted. Both sides trained their satellite eyes on Southeast Asia, waiting to see what would happen next. Washington told Beijing that it could not condone nuclear attack, no matter what the provocation; once the step to nuclear war is taken, the whole world could be engulfed in fireballs and mushroom clouds within hours. Moscow, at Hanoi's urging, sent a carefully worded message to Beijing, warning in much the same words that "breaching the nuclear threshold could bring on unforeseen and tragic consequences." Beijing's reply to Moscow was uncommonly curt: any attempt by the Soviet Union to intervene would be met by force, including nuclear weapons if necessary.

While the Supreme Soviet met in emergency session in Moscow, Nanning, the capital of Kwangsi Province, suddenly disappeared in a flash of nuclear fire. Unknown to either of the superpowers or to China, the leaders of Vietnam had foreseen that one possible response the Chinese might make against them would be to resort to nuclear weapons. So they had obtained, through means that are to this day not fully clear, a small cache of plutonium-fission bombs. Presumably they came from Red Army units stationed in Eastern Europe. At any rate, the sudden and thoroughgoing purge of Red Army units in East Germany and Czechoslovakia tends to reinforce this supposition. A flight of three Soviet-built MiG-27 "Flogger" fighter-bombers made a low-altitude supersonic dash to the Chinese city the night after the Haiphong bombing. The flight leader dropped his 50-kiloton bomb in the heart of the city, and all three planes sped home to safety as the center of Nanning was blasted out of existence. Nearly a third of the city's population — swollen to almost a million by the stream of refugees from the south — were killed or maimed by the explosion and resulting firestorm. Another hundred thousand received lethal doses of radiation and died within a few weeks.

Moscow immediately and very loudly denied any connection with the bombing. The Kremlin instituted an intense

Scenario Four: The Great Divide

investigation that very night, which quickly resulted in the purge of Red Army units in Eastern Europe. Washington expressed shock and horror and offered to mediate a truce between China and Vietnam. Both nations angrily rejected Moscow's excuses and Washington's offer. U.S. Navy intelligence, using data from surveillance satellites and sonar sensors emplaced on the floor of the Pacific Ocean, reported to the White House that all five of China's nuclear-missile submarines were cruising toward the west coast of the United States. The Chinese foreign minister informed the American ambassador in Beijing that this was a precautionary move only, intended to serve as a warning that if Washington should decide to join with the USSR in attacking China, the Chinese missiles would extract a just and terrible vengeance.

That is when it became clear to the American leadership that Beijing was ready to launch its land-based missiles at the Soviet Union. "Once that happens," a four-star general muttered, "anything can happen. Anything." The Soviet Premier spoke directly to the American President over the hot line, assuring him that Russia's nuclear missile force would be used against China only if the Chinese attacked first. He guaranteed that the Soviet Union had no intention of launching missiles against the United States or any of its allies.

To his everlasting credit, it was the American President who offered the first suggestion of mutual defense. "How many of the Chinese missiles do you think you can stop with your space-based defenses?" he asked, looking directly into the eyes of the Soviet leader on the picture screen of his videophone. "Not as many as we would like to," came the Russian's reply. "If the United States used its own space defenses to help you," asked the President, as his aides in the Oval Office went into various stages of apoplexy, "would the Soviet Union refrain from a counterstrike until we see how many of the Chinese missiles actually get through?"

The situation in the Kremlin office of the Premier must have been equally electrifying. The Premier glanced at the men around him, tough dedicated party members who were facing the threat of a hundred or more hydrogen bombs. No one said a word for many long moments. Then the Premier replied slowly, "Mr. President, we would be happy to have your help."

The President nodded. "And I presume that you will assist us if the Chinese should decide to launch their submarine-based missiles at the United States."

"Of course," said the Premier.

Plenty of high officials on both sides desperately opposed such cooperation, but the forces that pushed the two superpowers toward mutual defense were undeniable. For the Russians, there was a chance to stop most, if not all, of the Chinese missiles, *and to leave their own missile striking force intact.* If the defense was successful, the Soviet Union would still have its strategic missiles available to deal as it saw fit with China — or with America, if the tide of events should change. For the Americans, there was an opportunity to prevent a desperate Chinese attack from escalating into a general nuclear holocaust, and a chance to prevent the Chinese submarine-based missiles from launching a city-busting attack on the United States.

As Moscow and Washington issued a joint announcement of their mutual defense pledge, the American and Russian military establishments grudgingly began to exchange technical information on each other's orbital ABM systems. Mostly the data flow was from computer to computer. Very little of the information concerned the types or workings of the weapons aboard the satellites themselves; what was exchanged was information on targeting

Scenario Four: The Great Divide

systems, orbital positions, optimum target ranges, and reload times — just enough to make certain, within the few hours available, that there would be at least one satellite weapon available to attack each missile the Chinese might launch. Even though the data exchange was so limited, many military officers on both sides expressed their bitter disappointment and anger at being ordered to "spill our guts to the enemy."

Psychologists working for the National Security Council told the President's advisers that the joint Washington-Moscow announcement would have one of two effects on Beijing: either it would make the Chinese leaders cancel any plans they had to launch a nuclear strike on the Soviet Union, or it would infuriate them to the point where they would strike at both the Soviets and the United States. The one Chinese-American in the group, when asked his personal opinion, replied, "For them to back down now means not only a humiliating loss of face, but political and military defeat at the hands of Vietnam. They'll attack us both; they have no other choice. Their best hope now is Armageddon."

For half a day the world was paralyzed with fear. All the great cities of Europe and North America emptied themselves; thousands were killed in the mad, unplanned exodus. Riots and looting sprang up in the abandoned cities. Highways were clogged for hundreds of miles with frantic, wild-eyed people desperately seeking a refuge from the coming holocaust.

At precisely noon, Beijing time, the Chinese Premier closed his eyes and silently nodded to his defense minister. The minister picked up the red telephone at his elbow and spoke one word: "Yes."

Half an hour later, 232 missiles began rising from their silos in the mountain fastnesses of Xinjiang and Quinghai. Within seconds, American and Russian satellite weapons began slashing at them with their invisible, silent energy beams and hypervelocity missiles. Some of the warheads exploded, fireballs blossoming noiselessly high in the atmosphere or up above the ozone layer. Dozens of gigantic pulses of electromagnetic energy bubbled across the world, knocking out civil communications satellites, disrupting radio and telephone service all around the globe, shorting out electrical systems, causing the Northern Lights to flicker across the sky so brightly that they could be seen in full daylight as far south as Hawaii and Jamaica. Several warheads, knocked off course by the destruction of their booster rockets, exploded close to the ground and sent mushroom clouds filled with radioactive dust wafting ten miles high.

But none of the warheads got through to their targets in the Soviet Union. Not one Russian city was bombed. The Russian fields of missile silos escaped unscathed. Soviet citizens were killed: some of the "short falls" exploded in the birch forests or endless grassy steppes of Siberia. Although no towns were hit, fallout sprinkled lethal radiation over a score of villages, and plumes of gray smoke from forest and grass fires wafted halfway around the globe.

The submarines never launched their missiles. Apparently their commanders were under orders to attack the United States only if the Soviets launched a counterstrike at China.

It took many weeks before the world returned to some form of normality. A summit conference was quickly convened in Colombo, Sri Lanka, to deal with the war between Vietnam and China and the undeclared war between China and the Soviet Union. For the first time since Hitler's death, in 1945, Russians and Americans found themselves working toward common goals. Between them, they had the power to make the world obey.

Scenario Four: The Great Divide

The Peace of Colombo settled the Sino-Vietnam War by fiat: both sides would return to the status quo ante bellum, and an international patrol would be set up along the border between the two nations to guarantee the peace. The patrol, made up of American, Russian, Indian, and Malaysian troops, actually consisted more of machines than of men and women. Most of the human peacekeepers lived and worked aboard ships cruising in the Gulf of Tonkin, between Haiphong and the Chinese island of Hainan, out of sight of the Vietnamese and Chinese both. The border between the two nations was actually guarded by electronic sensors, backed up by drone surveillance planes. Smart missiles and tele-operated weapons systems, controlled by the technicians aboard the ships, prevented either side from massing troops or weaponry along the border.

The Colombo conference led, eventually, to the Treaty of Pisa, where — within sight of the famous Leaning Tower — the United States and Soviet Russia created the Multinational Space Defense Agency and invited all the other nations of the world to join it. A watershed in human history had been reached, a Great Divide that separates all earlier political thinking from that which followed. The threat of nuclear Armageddon had at last been lifted; the end of war itself was in sight.

Utopia had not been achieved. Russian political objectives still clashed with American political goals all around the world. Communists still worked to bring an end to capitalism, and capitalists still strove to drive Communism into the dust. But the means they used now were political and economic. Although both the superpowers still had enormous arsenals of nuclear and conventional weaponry, both had taken the first steps toward reducing those arsenals — and the huge budget expenditures that they demanded, year after year. Although the generals and admirals still insisted that placing total reliance for defense on a multinational force was arrant folly, although the armies and navies and nuclear strike forces did not disappear overnight, at least now there were voices speaking in the capitals of the world of lasting peace, of a gradual but consistent reduction in the level of arms, of shifting the emphasis in the national budget to areas that would help the twin causes of peace and prosperity.

Rome was not built in a day. Even God spent the best part of a week to create Eden. A world free from the fear of nuclear war — perhaps all war — would take many years. But it was on its way, at last.

19

Warfare Suppression

MUST THE WORLD come to the brink of nuclear holocaust before we take the steps necessary to prevent Armageddon? Must the deadly fireballs blossom again and thrust their mushroom stalks into the sky before we find the means to control nuclear war — and war of all kinds?

We are in a race against time. On the one side are the growing arsenals of the six (or is it seven? or eight?) nations that already possess nuclear weapons and the nations that are working to build their own. On the other is the technology of defense: ABM energy weapons and missiles, surveillance satellites, drone vehicles, and smart weaponry. The balance will be determined by the political decisions that we make, the decisions that will determine how these weapons and technologies will be used. We have about ten years in which to make up our minds; by the mid-1990s both the space-based defenses and the smart battlefield weapons will be ready for deployment. And perhaps ten or more nations will possess nuclear arms. That will be the moment of truth for the human race.

William Colby, former director of the Central Intelligence Agency, in a speech at the University of Virginia on 28 October 1983, said that nuclear weapons are "unusable," the arms race is "unwinnable," a unilateral weapons freeze is "unworkable," and the threat of nuclear war makes the world "unlivable." But he failed to tell his audience how to survive the perils of today and create a better world for tomorrow.

A similar warning has been sounded by Andrei Sakharov, the father of the Soviet hydrogen bomb and a noted Russian dissident, winner of the 1975 Nobel Peace Prize, who is now living in internal exile in Gorki because of his stands in favor of human rights and against nuclear weaponry. In a letter to Professor Sidney Drell of Stanford University, published in the summer 1983 issue of *Foreign Affairs*, Sakharov spoke chillingly of the consequences of nuclear war: "Mankind has *never* encountered anything even remotely resembling a large nuclear war in scale and horror."

Stating that hundreds of millions of human beings would be killed immediately by the blast, fire, and radiation of nuclear explosions, Sakharov went on, "No matter how appalling the direct consequences of nuclear explosions, we cannot exclude that the indirect effects will be even more substantial." He then described the onset of nuclear winter and the death of the human race:

> Continuous forest fires . . . could destroy the greater part of the planet's forests. The smoke . . . would destroy the transparency of the atmosphere. A night lasting many weeks would ensue . . . followed by a lack of oxygen in the atmosphere. . . . This factor alone . . . could destroy [all] life on the planet. . . .
>
> High-altitude wartime nuclear explosions in space . . . could possibly destroy or seriously damage the ozone layer protecting the earth from the sun's ultraviolet radiation. . . . If the maximal estimates are true, then this factor is sufficient to destroy life.
>
> Disruption of transportation and communication . . . disruption in the production and distribution of food, in water supply and sewage, in fuel and electric service, and in medicine and

clothing — all on a continentwide scale. . . . Sanitary conditions will revert to a medieval level. . . . It will be impossible . . . to provide medical assistance to the hundreds of millions who have been wounded, burned, or exposed to radiation.

Hunger and epidemics . . . could take more lives than the nuclear explosions directly. It is also not out of the question that . . . entirely new diseases could arise as the result of the radiation-caused mutation of viruses as well as especially dangerous forms of the old diseases, against which people and animals will have no immunity.

The consequences of nuclear war are so terrifying, so immense, that even those who force themselves to think hard about them cannot grasp all the implications. The end of civilization, the end of the human race, the end of all life on earth — these are words, concepts that we can formulate in our minds, but the visceral reality eludes us. It is too vast, too overwhelming, for us to *feel* as reality. Only late at night, in the darkness that swallows us as we drift toward sleep, does the nightmare fear truly grip our hearts. We could be the last human beings on earth. We could end it all tomorrow, tonight, in the next hour.

As I write this, the United States is fulfilling its pledge to NATO to install cruise missiles and Pershing IIs in West Germany, Britain, and Italy. The negotiations on Euromissiles have been broken off, and the Soviet leadership threatens to install Russian cruise missiles to counter the American weapons. Meanwhile, the Soviets are continuing to install SS-20s in Russia's European region. The START negotiations on strategic weapons appear to be stalemated; no one expects significant progress there until well after the presidential election of 1984, if then. The U.S. defense budget, soaring toward $300 billion per year, includes plans to spend more than $12 billion over the next five years on research and development for a space-based defense system. There is no way to tell how much the Soviets are spending on defense, although estimates range as high as 12 to 20 percent of their gross national product. U.S. defense outlays have remained below 6 percent of GNP since the Korean War. The Soviet GNP, however, is only about two-thirds that of the United States.

The history of the past four decades has been the history of the arms race between the United States and Soviet Russia. In 1945, the United States enjoyed sole possession of nuclear weapons, but that did not prevent the Soviets from establishing their long-sought suzerainty over Eastern Europe, a goal of Russian foreign policy since the time of Peter the Great. By 1949 the USSR had created its own nuclear bomb and was well along the road to building long-range missiles that could carry their warheads to the American heartland. Both superpowers developed hydrogen bombs at about the same time, in the mid-1950s. In the 1960s the race was to produce the fleets of ballistic missiles that became the backbone of Mutual Assured Destruction. A spinoff of this competition was the widely heralded space race, which sent American astronauts to the moon and saw unmanned spacecraft from both superpowers begin the exploration of the other worlds of the solar system. Despite sincere efforts to negotiate limits to the weapons race, the policy of MAD ensured that both nations would strive to build deadlier and deadlier nuclear "deterrents," and an alphabet-soup chain reaction led from ABM to MIRV to ASAT to MX.

For almost forty years the United States has perceived that the position of world dominance it enjoyed in 1945 has been steadily eroding. For forty years the Soviet Union has perceived that its survival depends on catching up with the Americans and surpassing them wherever possible. Although the basic nature of MAD policy *requires* that the two superpowers have roughly equal strategic arsenals, both sides constantly see their enemy as being "ahead." What one nation sees as a desperate effort to "catch up," the other sees as a deliberate attempt to "escalate the arms race." These phrases are seldom used cynically; they are not mere political rhetoric. With 20,000 megatons of nuclear power already built into the world's arsenals, the fears are very real.

Now, after forty years of constant escalation in offensive arms, comes the possibility of defense. The technologies that have led to the danger of nuclear devastation also offer us the hope of ending war forever: as Shakespeare put it, "Out of this nettle, danger, we pluck this flower, safety." Assuming that the technologies of energy weapons and automated battlefields will become workable in another decade or so, should we proceed toward space-based defenses and warfare suppression? Should we encourage Benford's "defensive arms race" or Ullman's internationalized ABM system?

Senator Larry Pressler, Republican of South Dakota, is opposed to the Star Wars option.

> By moving the arms race into space [he says], the danger of nuclear war on earth will not decrease. On the contrary, a space arms race would not only upset the delicate balance that prevents such terror from occurring but would also increase the risk of a total cosmic conflict. . . . In the short run, the enormous financial cost of space weaponry would impose a staggering tax burden on the average American and would also reduce the amount that our government would be able to spend on conventional military programs. In the long run, a space arms race would drain the world of resources needed for countless other purposes. . . . America's interests are best served if the skies are clear of deadly weapons. Let's abandon these fortresses in the sky.

Peter A. Clausen, of the Union of Concerned Scientists, also weighs in against space-based defenses, primarily because he believes the Star Wars option works against meaningful arms control agreements and could make arms limitation negotiations "irrelevant by undercutting any transition toward fewer and smaller nuclear missiles."

Clausen claims that

> the vision of a defense against missiles appeals to two groups: idealists . . . who want to remove the threat of nuclear war, and strategists (who seem to have shaped the proposed program) looking for a new route to military superiority over the Soviet Union. The hopes of both are destined to be disappointed. The most probable result of a major missile defense program would be to doom existing arms control prospects while driving the arms race into more expensive and destabilizing paths, with less security for both nations.

Clausen, Senator Pressler, and most of the others who oppose the idea of space-based defenses express three major fears:

1. Because an orbital ABM system might work, it would degrade the mutuality in MAD, and thus destabilize relations with the Soviet Union.
2. If the Soviet Union fears that a space-based defense system would encourage the United States to launch a pre-emptive attack, the Russians would either try to destroy the satellites before they become functional or launch a nuclear strike on the United States before the ABM system goes into operation.
3. The costs of a space-based defense system would be staggering; every dollar spent on Star Wars would be a dollar that could not be spent elsewhere in the defense budget.

To these objections, a fourth could be added. If the United States embarks on a multibillion-dollar program to develop a space-based defense system, the Soviet Union might well do the same, thus escalating the arms race into orbital space.

All of these objections, valid or not, are based on an underlying assumption: that the United States *unilaterally* proceeds with development of a space-based defense system.

What if the United States invited its allies and the nations of the third world to join in the program? After all, President Reagan mentioned that possibility in his 23 March 1983 speech. Certainly the NATO nations and other allies of the West would expect to be sheltered under the umbrella of an American space-based ABM system. Why not extend the offer to the third world nations, as proof of our sincerity? Would they not be willing to share some of the costs? To put some of their scientists and engineers to work on the program? At the very least, they could serve as a sort of inspection team, guaranteeing that the systems being built are defensive and intended to protect all the peoples of the earth.

What if the United States invited *all* the nations of the world to join in the program — especially the Soviet Union and its Warsaw Pact allies? As Ullman pointed out, this would mean sharing much technical information that is now classified secret by the Department of Defense. Hardly anyone in the Pentagon would agree to such a move; everyone would blanch at the thought of sharing our space technology and energy-beam developments with the dreaded and detested Russians.

Yet Jerry Pournelle, who flatly describes the leaders in the Kremlin as "a gang of aged homicidal maniacs," has no qualms about sharing the technology of space-based defenses with the Soviets. "Shields are shields," he told me. "If the Russians put up defensive satellites, that doesn't hurt me. I don't want to burn Russian schoolgirls; I want us to move to a policy of Assured Survival."

But, like MAD, Assured Survival must be *Mutual* if it is to work. The only way to make the transition from offensive weaponry to defensive is to make certain that neither of the superpowers perceives its interests to be threatened by such defenses. The way to assure mutual survival is to make the development of space-based defenses a mutual effort — indeed, it must become a worldwide program. Orbital ABM satellites are *global* entities, by their very nature. If they can protect one nation, they can protect every nation. If they threaten one group of people, they threaten all the world's people.

MIT's Eric Drexler has said:

> If . . . an automatic BMD system could recognize and fire upon *any* flight of missiles large enough to pose a first-strike threat, and if both sides knew this, and both could verify that the BMD system did not know the difference between the United States and USSR, and that it would ignore satellites, shuttles, and space stations, then the stabilizing advantages of BMD could be had without unilateral advantages or a threat to peaceful space activities.

It takes only that one step in our thinking, the step from a national view to an international one, from chauvinism to global consciousness, to see that the technology of space-based defense leads inevitably to a powerful change in our political outlook. Space-based defenses *might* work if one of the superpowers puts them in place. They will certainly work if they are established in orbit by all the nations of the world, as a defense for the whole human race against any nation that dares to launch a nuclear missile attack against any other nation.

Such ABM satellites would not make war impossible; they would not even be able to interfere with nuclear attacks that are carried out by low-flying planes, cruise missiles, or conventional artillery. Certainly they would be useless against terrorists or saboteurs who sneak suitcase-size nuclear weapons into a country. But they would eliminate the biggest threat facing the human race: total annihilation, resulting from the launch of thousands of unstoppable missiles that carry megatons of nuclear devastation in their warheads.

Is it reasonable to expect that the Soviet Union, if invited, would join such an international effort? Perhaps not. I suspect that the Soviets would try to have their cake and eat it too: they would want to send "observers" to participate in the Western development program while they work in secrecy on their own. As Arthur Clarke might say, "So what?" The intent of the Western program would be to convince the Russians that, as President Reagan said, "Would it not be better to save lives than to avenge them? Are we not capable of demonstrating our peaceful intentions by applying our abilities and our ingenuity to achieving a truly lasting stability?"

Even if the Soviets built their own space-based defense network while the West built theirs, the worst that would happen is that the world has *two* ABM systems in orbit — and twice as much military money siphoned away from offensive armaments. True, the weapons in orbit could be turned against other satellites, either the other side's ABM battle stations or the surveillance and communications satellites that orbit the globe. But that would be unlikely, as long as both East and West had roughly comparable capabilities in orbit. Again, mutuality leads to stability; it is imbalance that encourages aggression.

And the Soviets might very likely welcome defensive systems that reduce or eliminate the effectiveness of nuclear-tipped missiles. Since they have a massive preponderance of conventional arms in relation to the West, the leaders in the Kremlin might well see Mutual Assured Survival as being to their advantage.

Two policing systems in orbit? Redundant, but perhaps workable. Inevitably, as time proceeds and the threat of nuclear missile attack becomes more remote, the tensions of today would begin to ease. Slowly the people of the world will learn to accept Mutual Assured Survival, and the two space-based defense systems will gradually merge into one, as the technicians operating the two systems and the political leaders of the two nations learn to trust each other.

In the meantime, much the same kind of politicotechnical symbiosis could lead to the development of automated battlefield systems and the multinational political arrangements to control them. Surveillance and communications satellites (like the U.N.'s ISMA and Arthur Clarke's Peacesats), together with smart weapons and tele-operated drone vehicles, could eventually remove human soldiers from the battlefield. If the superpowers can cooperate enough in space, however grudgingly, to remove the threat of nuclear missile attack, they may learn to cooperate on earth enough to make conventional war unlikely.

The scenario of warfare suppression may follow this track. The Western allies, together with key third world nations, such as India, create a multinational space-based defense system. The Soviets follow suit with their own orbital ABM satellites. As East-West tensions ease, the two systems function more and more cooperatively. Capitalizing on this momentum, the West encourages the third world nations to create an International Peacekeeping Force consisting of peacekeeping satellites, smart weaponry, and a small cadre of men and women to direct the system. No nation is required to disarm; the IPF's mandate is merely to prevent aggression across international borders. If a nation wishes to maintain its own defenses, the IPF has no objection.

Gradually, though, many nations begin to realize that they can ease the armaments burdens they carry by reducing the size of their national defense systems and relying on the IPF to protect them from aggression. Undoubtedly, the United States and the Soviet Union would be the last nations to do so. (Or maybe not. Switzerland, which has for centuries followed a policy of heavily armed neutrality, might well be the very last.)

It takes a fundamental shift in outlook to accept the idea of an International Peacekeeping Force, armed with everything from orbital ABM lasers to automated battlefield missiles, replacing the armies, navies, air and space forces of today. Would the average American be willing to disband the U.S. military establishment and leave the defense of the nation to an organization made up largely of foreigners? Would the average Russian?

How would the IPF be constituted? How much say would the United States, or any other single nation, have in its operation? Is not the IPF merely a first step toward a world government? Could not the Communists infiltrate the IPF and use it as a front for world domination?

How would the world change if the $600 billion per year that the nations spend on armaments were diverted, in large part, to more peaceful pursuits, such as food production, education, industrial development, housing, medical care, and sanitation? Much of the land mass of Europe and North America, buried under mile-thick glaciers during the Ice Age, is today slowly rising, lifting itself, now that the burden of the ice has melted from the soil. How will the human spirit rise, rebound, once the crushing burden of armaments has been lifted from its shoulders? How will civilization advance, once we can devote our best efforts to the pursuits of peace?

There are no answers to those questions, nor to a hundred more like them. Not yet. The answers have to be worked out, patiently, carefully, over the years to come.

The point to remember, though, is this: the tools for warfare suppression are at our fingertips. Within a decade, the human race will have the technological means to prevent nuclear missile attack and to stop ground or sea warfare. We have ten years to decide on how to use these technologies. The step from today's bipolarized world of Mutual Assured Destruction to a peaceful world of Mutual

Assured Survival and warfare suppression is neither simple nor small. It is truly a giant leap for mankind. But even the longest journey begins with a single step.

As John F. Kennedy said in his inaugural address, "All this will not be finished in the first hundred days. Nor will it be finished in the first thousand days, nor in the life of this administration, nor even perhaps in our own lifetime on this planet. But let us begin."

Let us begin. Now.

Epilogue: The Athenian Ideal

The biggest cause of trouble in the world today is that the stupid people are so sure about things and the intelligent folks are so full of doubts.

— Bertrand Russell

It is a quiet Sunday morning as I write this. Autumn has transformed the trees outside my Connecticut window into breathtaking splendors of gold and red. The sun is shining warmly, even though the air tingles with a brisk foretaste of the winter that is to come.

I watch a neighbor walking by. A car swings around the curve in the road in front of my house. And I think how beautiful this place is, how beautiful this world is, and how precious is the peace and freedom that we often take for granted.

During the course of my life I have been fortunate enough to make friends in many places around the world, and I think of them now, in London and Moscow, in Athens and Tahiti, in Buenos Aires and Paris and Dublin and Colombo and a hundred places across the United States. Each of them loves his or her home as much as I love mine. Each of them wants peace and the right to live life as he or she sees fit. I don't know anyone who wants war.

Some of my friends are military officers; others are officials in several different governments. None of them looks forward to fighting a war. I don't believe that there are men in the Pentagon, the White House, the Kremlin, or anywhere else on earth who are actually planning to start a nuclear holocaust.

Yet I know just as well that some of my friends work on plans to fight a war if it should start. Some of them, if ordered to do so, would launch nuclear-armed missiles. They do not enjoy bearing that responsibility, yet they accept it because they are certain that the independence of their nation depends on such determination.

A few of my friends are beginning to realize that the tools for peacekeeping are at hand. They understand that today's technology of computers, lasers, and satellites offers us a chance to create an International Peacekeeping Force that will suppress the war-making abilities of nations. They perceive that war may, at long, long last, become preventable. They see that the terror of Mutual Assured Destruction can be replaced by the armed peacekeeping forces of Mutual Assured Survival.

I have tried to share this understanding with you. I have tried to show how the technology of warfare suppression can be brought into being. But the technology by itself is meaningless. Only the moral decision to stop war, and the political action that must flow from such a decision, can actually eliminate warfare from the human scene.

It is one of the great paradoxes of history that the terrible destructive power of nuclear weapons has made a major war less likely than it once was. Yet the other side of that coin is that if a nuclear war does come, it may well destroy all human life.

Now we stand at the edge of another such paradox. By placing weapons in space, and developing highly automated computer-driven systems that can replace human soldiers, we are producing the tools that can be employed to prevent warfare altogether. We can create a highly automated international police force that will prevent war from starting. This generation of human beings, you and I, have the opportunity to put an end to war. The gamble we are taking is a huge one; the stakes are the entire future of humanity. Do we trust our powers of creation as much as we now trust our powers of destruction?

Whether you look at the problem of war from the plane of religious morality or from the level of practical politics, one fact is eminently clear: either we end war or war will end us.

Orbital weaponry will be built; have no doubt of it. Laser-armed satellites will be patrolling the skies over our heads whether we like it or not. The smart bombs of today will give rise to brilliant drone weapons and tele-operated fighting systems tomorrow. Will we have the political sense and the moral courage to create a true International Peacekeeping Force, to yield enough of our national sovereignty to make war impossible? Or will the extension of weaponry into space and the advent of computerized battlefields merely bring the day of Armageddon closer?

We still have time to make a choice. But before we choose, we must think. We must seek out the facts and ponder them carefully.

The ancient Greeks worshiped two gods of war or, rather, a god and a goddess. Ares (whom the Romans called Mars) was the aggressive, bloodthirsty god of violence who liked nothing better than to see men fighting and killing each other. Athena, who was not born of woman but sprang full grown from the brow of Zeus, bearing shield and spear, was originally a battle goddess, but as time evolved gray-eyed Athena also became the goddess of wisdom, of learning, of civilization and democracy. Her symbol was the owl, and ancient Athens became her special city. She remained a warrior goddess, but she represented to the Greeks the craft of defensive war, of strategy and planning, of careful preparations to minimize bloodshed.

It is time that we, with this generation of awesomely terrible weapons in our hands, turn away from bloody Ares and his battle lust and turn toward wise Athena. It is time that we begin the long, difficult road toward a world of peace, a world that is freed from the crushing burden of armaments, a world in which our children and our children's children will be free at last of the fear of war. We have at our fingertips the tools to fulfill the biblical prophecy:

And they shall beat their swords into plowshares, and their spears into pruning hooks: nation shall not lift up sword against nation, neither shall they learn war anymore.

Appendix: Space Weapons Bibliography

Appendix: Space Weapons

THE WEAPONS being developed for use in space can be regarded as "black boxes"; that is, it is sufficient to know what goes into them in the way of power or fuel requirements and what comes out of them as energy or projectiles. For most readers, it is not necessary to understand what goes on inside the black box, how the weapon works. But for those who are curious about the inner workings of these exotic new kinds of weapons, here is a thumbnail sketch of their fundamental principles.

The space-based weapons discussed in this book fall into two basic categories: energy-beam weapons and kinetic-kill weapons. Each category has two types of weapons. Energy-beam weapons, often referred to as directed energy weapons (DEW), consist of lasers and particle-beam accelerators. Kinetic-kill weapons include missiles and electrically powered rail guns.

Lasers

The word *laser* is an acronym of light amplification by stimulated emission of radiation.

In a laser, a working medium is made to emit an intense beam of light. The working medium can be a solid, a liquid, or a gas. The first lasers used solids, such as artificial rubies. The lasers being developed for space weapons use gases, like carbon dioxide or hydrogen fluoride, because gas lasers can produce much more power than solids.

Light is energy, pure and simple. Well, maybe not so simple. Twentieth-century physicists discovered that light can be considered a wave of energy or a stream of massless particles, which they dubbed *photons*. Or both. Visible light is a small slice of the huge spectrum of electromagnetic energy, a spectrum that includes radio waves, infrared energy, visible light, ultraviolet, x rays, and gamma rays. Radio waves are the longest in wavelength and the lowest in frequency; some radio waves are miles long. The waves of visible light, in contrast, are less than ten millionths of an inch long. The energy content of a photon depends on its wavelength. Photons of the long-wavelength radio waves, for example, pack much less energy than photons of visible light. Ultraviolet light is more energetic than visible light, x rays more energetic than ultraviolet, and gamma rays the most energetic of them all.

A laser emits energy in the form of an intense beam of light — or, rather, an intense beam of electromagnetic energy—which can range from infrared through visible light to ultraviolet. X-ray lasers are also being developed. No one has produced a gamma ray laser. Yet. (Radio-wavelength devices called *masers* — for microwave amplification by stimulated emission of radiation — actually preceded the development of lasers. But that's another story, and does not concern us here.)

Regardless of its wavelength, the energy that comes out of a laser is very different from the light emitted by ordinary lamps. Laser light is very intense, it is emitted in a narrow beam, it is monochromatic, and it is coherent.

Intensity. The light produced by lasers is much brighter than the sun. The sun emits some 7000 watts (7 kilowatts) of light energy from each square centimeter of its surface — equivalent to the light output of seventy 100-watt bulbs, coming out of an area the size of a postage stamp. Lasers can produce pulses of light of more than a billion watts per square centimeter. Laser beams have been focused down to intensities of more than 10 billion watts per square centimeter.

Directionality. Lasers emit narrow beams of light that spread very little, even over vast distances. Lasers of quite low power, merely a few watts, have bounced their beams off the moon, spreading to little more than a mile's diameter after a trip of a quarter million miles. This directionality means that comparatively little of a laser's beam is wasted; if the laser is used as a weapon, most of its energy would get to its target.

Monochromaticity and Coherence. Ordinary sources of light, such as candles, electric lamps, or the sun, emit light at many different wavelengths simultaneously. This is called *white light.* Lasers emit light at a single wavelength, a single pure color of such richness that laser light shows have become popular entertainments. The difference is rather like the difference in sound between a symphony orchestra as it tunes up and the pure note of a solo violin. Laser light is also *coherent:* all the waves coming out of the laser are lined up in near-perfect step. Ordinary light sources produce light waves of many different wavelengths, mixed together. Lasers produce light waves that are more precisely spaced and aligned than a marching squad of West Point cadets.

What happens inside a laser to produce such extraordinary results?

Atoms give off light quite naturally. The sun and stars shine. Flames glow. Electric lights chase away the shadows.

In each of these light sources the atoms are excited by some energy source: thermonuclear fusion in the sun and stars, chemical reactions in a flame, electrical energy in artificial lights. When an atom absorbs energy, physicists say it is excited. But an atom's excitement does not last long under natural circumstances. Within a tiny fraction of a second, an excited atom will give off exactly the same amount of energy that it just absorbed. It emits this energy in the form of a photon, and the photon has precisely the same wavelength (or energy content) as the energy that the atom had absorbed a few microseconds earlier.

Ordinary light sources are not monochromatic or coherent, as lasers are, because the atoms within them are absorbing and emitting photons of energy randomly, haphazardly. In a laser, conditions are deliberately arranged so that the atoms of the working medium absorb energy and then release it in a very strictly controlled manner. A laser compares to ordinary light sources, as far as the allowed behavior of the atoms is concerned, rather like a maximum-security prison compares to a surburban shopping mall.

In the laser, the atoms of the working medium are excited by some input of energy. Then they are forced to remain in this excited state for a much longer time than they would under natural conditions. This condition is called a *population inversion*. It means that a large percentage of the population of atoms is "hanging up" in the excited state — the inverse of the normal situation, where only a tiny fraction of the atoms would be excited at any given moment.

The excited state is maintained for a certain period of time (a fraction of a second can be a very long time in atomic processes), and then the excited atoms are made to dump their stored energy. A cascade of photons flows from the laser, all of them at the same wavelength. A beam emerges, far brighter than the sun, as directional as an arrow fired by Apollo, monochromatic and coherent.

That beam can carry enormous energy. Although the earliest lasers emitted pulses of light of only a few watts' overall power, the breakthrough to gasdynamic lasers has led to lasers in the multimegawatt range.

There are three main types of high-power lasers now being developed. They differ mainly in the source of energy they use to produce the population inversion of excited atoms within the laser cavity.

The original gasdynamic laser produced energy from combustion. The earliest gasdynamic laser burned cyanogen and oxygen, and the laboratory was liberally stocked with gas masks: cyanogen is deadly. Burning cyanogen produced a hot mixture of carbon dioxide, nitrogen, and traces of other gases. The carbon dioxide was the gasdynamic laser's working medium. The laser emitted infrared energy.

Electrical energy can also be used to excite a gas laser. The earliest carbon dioxide lasers were electrically driven, and today the most efficient high-power lasers use electrical excitation in one form or another to produce the population inversion. Efficiencies of 20 to 30 percent are theoretically possible, though most lasers — including the electrically driven types — are not yet that efficient.

The wavelength of the output beam of a laser depends on its working medium, not its energy source. Thus, electrically powered lasers come in as many different varieties as the working gases they use. The highest power outputs come from carbon dioxide lasers, and they emit in the infrared range. But electrical lasers using other gases emit in other wavelengths.

It is also possible to use the energy generated by chemical reactions to produce laser action. The most common chemical laser mixes hydrogen and fluorine to produce a powerful beam of infrared energy. In some cases, deuterium is used instead of ordinary hydrogen to move the output beam to a wavelength that is not absorbed by air as much as the slightly different wavelength produced by the hydrogen-fluorine reaction. Deuterium is an isotope of hydrogen; it is chemically similar, but where hydrogen has merely a single proton in the nucleus of its atom, deuterium has a proton and a neutron. When deuterium takes the place of ordinary hydrogen in water, the stuff is called "heavy water." It is just as drinkable as ordinary water, although it tends to muddle up one's sense of balance when it replaces the ordinary water in the semicircular canals of the inner ear.

X-Ray Laser. Perhaps no secret weapon development program has been better publicized than the x-ray laser. The brainchild of Edward Teller and his staff at Lawrence Livermore National Laboratory, in California, the x-ray laser offers a great advantage over other laser weapons — and presents a great disadvantage.

The advantage is its wavelength. X rays pack much more energy than lower-wavelength visible and infrared laser beams. An x-ray laser would be much more lethal.

The disadvantage is that the x-ray laser, as currently envisioned, would be powered by the explosion of a small nuclear bomb. To Teller, father of the U.S. hydrogen bomb, a nuclear device is something to be used, not something to be feared. He calls the combination of bomb and laser a third-generation nuclear device. The first generation was the fission bomb, the kind that destroyed Hiroshima and Nagasaki. The second generation is the hydrogen bomb, a thousand times more powerful.

The third-generation nuclear device would transform the energy released by a small nuclear explosion into *directed energy,* a beam of x rays emitted by a laser. Just how this is accomplished is still shrouded in military secrecy, but the general outlines may be guessed at. A 1-kiloton explosion (about one twentieth of the Hiroshima bomb) releases the equivalent of one trillion kilowatt-hours of electrical energy. Even assuming fairly inefficient means of converting the energy of a small nuclear explosion into electricity, a very powerful electrically driven laser can be built with a small nuclear bomb as its primary energy source. The electricity would energize a working medium, presumably a gas, and excite the atoms in it to such an intense pitch that they emit x radiation.

Regardless of their wavelengths, laser beams can cross long distances in the vacuum of space, where there is no air to absorb their energy. They move with the speed of light, 186,000 miles per second in vacuum; nothing in the universe goes faster. The beams also travel in straight lines, undeflected by electrical or magnetic fields, or even by the gravitational pull of the earth. (Light can be deflected by *very* massive gravitational forces, such as the pull of the sun, but such deflections are so small that it takes precise astronomical equipment to measure them. In the case of laser beams fired from satellites orbiting close to the earth, there is no practical effect from the sun.)

Laser beams are, in fact, both the fastest gun in town and the equivalent of the old science fiction death ray. The focusing of several kilowatts or megawatts of energy per square centimeter will cause damage to any target similar to the kind of damage done by Buck Rogers' "disintegrator" ray gun.

The basic "kill mechanism" of a laser beam is to heat the target's surface, quickly and to very high temperatures. The skin of a missile, for example, can be boiled away by the searing finger of a laser beam, which can punch a hole in the missile's skin in a second or less. One possible problem with this, however, is that the boiling metal may create a cloud that will tend to absorb the energy of the incoming beam. The latest analyses indicate that the cloud would blow away before it interfered, to any significant extent, with the incoming laser energy. Even if this is not the case, the beam could be pulsed so that the cloud created by the first pulse of energy certainly dissipates before the next pulse arrives. The pulses could be thousandths of a second long, or even less.

Very high energy pulses could also damage or destroy a target by mechanical shock, literally shaking the target apart. A train of sufficiently energetic pulses could rattle a satellite or missile to pieces.

Particle Beams

Devices that accelerate subatomic particles are hardly new. For generations, physicists have probed the inner workings of the atomic nucleus with machines that accelerate subatomic particles such as protons or electrons. The particles are fired into atoms and split their nuclei apart so that the physicists can see what's going on inside them. These atom smashers are called cyclotrons, synchrotrons, bevatron, tevatron, SLAC, to name a few.

The basics of particle accelerators are simple, although the engineering and operation of these machines are not. Basically, they all take advantage of the fact that most subatomic particles carry electrical charge. The electron carries a negative charge. The proton, a positive charge. These particles are *minuscule*, by the way: it takes 350 trillion trillion electrons to equal one ounce, and a trillion electrons placed alongside each other would barely cross the width of your little finger. Protons are 1837 times heavier than electrons, but that still means that an individual proton weighs on the order of five billion billion billionths of an ounce.

Small though they are, electrons and protons are electrically charged, and therefore can be accelerated to extremely high velocities if they are pushed or pulled —

or both — by electromagnetic forces. Particle accelerators are, in principle, little more than sets of very powerful magnets and electric generators. They can accelerate particles virtually up to the ultimate speed limit of the universe, the velocity of light — 186,000 miles per second.

Even the tiniest particle has considerable kinetic energy at that velocity. A stream of electrons or protons blazing along at nearly the speed of light can inflict damage on a target just as a bolt of lightning does. Particle accelerators with energy inputs of a trillion volts have been built on the ground, for research. If such a powerful accelerator were placed in orbit, the particle beam coming from it could be a very effective weapon against satellites or missiles.

But the beam would have to be neutralized, electrically, before it could be useful as a weapon. A beam of electrons or protons would be immediately deflected by the earth's magnetic field. The negative electrons, or positive protons, would be bent by the magnetic field, rather than traveling in a straight line from accelerator to target. So a neutral beam is needed. The most commonly discussed system would accelerate protons to a speed almost that of light and then pass them through a cloud of electrons as they exit the accelerator. The comparatively massive protons would not be impeded by the lighter electrons, and the electrons would be grabbed by electrical attraction and attached to the protons as they sped past, much like a waiting sack of mail is picked up by a speeding express train's extended hook.

The neutral beam would then travel straight to its target and deliver a massive jolt of energy to it. Such a particle beam would not be absorbed by a cloud of gas, as a laser beam would be. Nor would it be reflected by a shiny surface or absorbed by an ablative coating. It could penetrate the metal skin of a missile, or even the "hardened" heat shield of a re-entry warhead, within microseconds. The beam could shock-heat the inner workings of a nuclear bomb, destroying its electronic controls or damaging the triggering mechanism so badly that the bomb would not detonate.

The kill mechanics of particle beams seem to offer great advantages over those of lasers, which must heat their targets relatively slowly. Even though everything happens in thousandths or even millionths of a second, the laser is to the particle beam as a blowtorch is to a battering ram.

But a particle-beam accelerator requires enormous electrical energy. It may be possible to "drive" an accelerator in space with a nuclear reactor, which generates electricity that is stored in capacitor banks and then released in a microsecond surge of a trillion volts or more. Alternatively, a small nuclear bomb, akin to the third-generation devices proposed for driving x-ray lasers, may be needed to power the accelerator. It seems clear that solarvoltaic panels, fuel cells, and other less intense electrical power systems will not be sufficient to drive an orbiting particle accelerator.

Like lasers, space-based particle-beam accelerators will not be a threat to targets on the surface of the earth. Their beams would be absorbed and scattered by the atmosphere long before they penetrated close to the ground. As the physicist (and science fiction author) Gregory Benford put it, "No sky-high battle station will ever pound a city into submission. But they can be planetary policemen, spreading a shield over the upper atmosphere."

Missiles

Set a thief to catch a thief. Use a missile to shoot down a missile. In the 1950s and sixties, both the United States and Soviet Russia spent tens of billions of dollars on ABM missile systems. They were designed to be launched from the ground and to use relatively small nuclear bombs to destroy the incoming hydrogen bomb warheads of the attacking force.

The missiles proposed for use in space-based defenses do not require nuclear warheads. Since they would be launched from orbiting satellites moments after the attacking ballistic missiles are launched, they could be guided to the attacking missiles while the rocket boosters are still blazing, or during the half-hour coasting flight before the warheads re-enter the atmosphere. They could be guided accurately enough, it is believed, to destroy their targets either with conventional explosives or by direct impact — the kinetic-energy kill technique.

The miniature homing vehicle of the U.S. Air Force's ASAT missile, currently under test, is a reasonable example of what a satellite-launched ABM missile might look like. The MHV weighs thirty-five pounds, is thirteen inches long, and twelve inches in diameter. Guided by miniaturized infrared sensors and an on-board computer, and maneuvered by fifty-six small steering thrusters, the MHV flies into the target satellite and destroys it by force of impact. Much the same kind of MHV could be used against a ballistic missile or a hydrogen bomb–carrying warhead. In such a case, a good deal of the impact force would be provided by the target's own momentum: ballistic missiles travel at speeds of many thousands of miles per hour.

The High Frontier system described by General Daniel O. Graham also uses small ABM missiles, carried aboard a fleet of 432 satellite "trucks." Each truck bears forty to forty-five such missiles. These missiles can accelerate to more than two thousand miles per hour, relative to the velocity of the satellite truck. The missiles consist only of propulsion rocket and warhead; they are guided by a computer on the truck and a radio command link. The warhead is a nonexplosive kinetic-kill device.

Rail Guns

Picture an electric catapult hanging in the sky. It is a long, narrow piece of hardware, the length of several football fields. At one end are a bulky, massive electrical power generator and associated power conditioning equipment. From this leads a long, slim metallic barrel, almost like the barrel of some giant's fowling piece. At the end of the barrel is another cluster of electrical equipment.

The mass of this complex piece of hardware is more than a hundred tons. Its job is to fire darts that weigh only a few pounds. But it fires them exceedingly fast: 2.5 million centimeters per second, as they leave the barrel. The darts, or *flechets*, as they are sometimes called, go from zero to better than fifty thousand miles per hour in considerably less than one second.

When those darts hit an object, especially an object that is itself hurtling through space, as a missile or warhead does, the impact is shattering. Kinetic kill at its most energetic.

The technology for rail guns, which is what these electric catapults are called, evolved out of peaceful research. For years, engineers have sought to build railroad trains that are levitated on magnetic fields, needing no wheels and capable of speeds comparable to a jet airplane. Even earlier, in 1950, Arthur C. Clarke described the idea for an electrically powered catapult to launch cargo and people off the surface of the moon. Gerard K. O'Neill expanded on this idea in his book *The High Frontier* (not to be confused with General Graham's High Frontier study and report). O'Neill and his Princeton students have built laboratory models of the device, which they call a "mass driver." It is a close cousin to the linear particle accelerator used by particle physicists; indeed, O'Neill's early interest in such devices stemmed from his work in particle physics.

Like the particle accelerator, the rail gun uses electromagnetic forces to move an object. But instead of accelerating subatomic electrons or protons, the rail gun accelerates macroscopic objects. In its incarnation as O'Neill's mass driver, it is intended to fling payloads of lunar ores off the surface of the moon, to feed space construction projects. As a potential space weapon, it fires darts — hard enough to destroy ballistic missiles and their warheads.

Also like the particle-beam weapon, the rail gun requires a massive electrical power source. And it must aim its projectiles with unerring accuracy, since the darts are too small to carry guidance and propulsion equipment. Once out of the barrel, the darts fly to their targets in presumably straight lines.

Bibliography

Arbatov, Georgi, and Willem Oltmans. *The Soviet Viewpoint.* New York: Dodd, Mead, 1983.

Baker, David. *The Shape of Wars to Come.* New York: Stein and Day, 1982.

Bova, Ben. *The High Road.* Boston: Houghton Mifflin, 1981.

von Braun, Werner, and Frederick I. Ordway. *History of Rocketry and Space Travel* (third edition). New York: Crowell, 1975.

Canan, James. *War in Space.* New York: Harper & Row, 1982.

Cockburn, Andrew. *The Threat: Inside the Soviet Military Machine.* New York: Random House, 1983.

Cohen, Sam. *The Truth about the Neutron Bomb.* New York: William Morrow, 1983.

Collins, John M. *U.S.–Soviet Military Balance.* New York: McGraw-Hill, 1980.

Divine, Robert A. *Eisenhower and the Cold War.* New York: Oxford University Press, 1981.

Dunnigan, James F. *How to Make War.* New York: Morrow, 1982.

Freedman, Lawrence. *The Evolution of Nuclear Strategy.* New York: St. Martin's Press, 1982.

Galbraith, John Kenneth. *The Anatomy of Power.* Boston: Houghton Mifflin, 1983.

Goldman, Marshall I. *USSR in Crisis.* New York: Norton, 1983.

Graham, Daniel. *High Frontier.* New York: Tom Doherty Associates, 1983.

Grey, Jerry. *Beachheads in Space.* New York: Macmillan Publishing Co., 1983.

Hackett, Sir John. *The Third World War: The Untold Story.* New York: Macmillan, 1982.

Bibliography

Hackett, Sir John, et al. *The Third World War: August 1985.* New York: Macmillan, 1978.

Hecht, Jeff. *Beam Weapons: The Next Arms Race.* New York: Plenum Publishing, 1983.

Hecht, Jeff, and Dick Teresi. *Laser: Supertool of the 1980s.* New York: Ticknor & Fields, 1982.

Kaplan, Fred. *The Wizards of Armageddon.* New York: Simon & Schuster, 1983.

Karas, Thomas. *The New High Ground.* New York: Simon & Schuster, 1983.

Kennan, George F. *The Nuclear Delusion.* New York: Pantheon, 1983.

Luttwak, Edward N. *The Grand Strategy of the Soviet Union.* New York: St. Martin's Press, 1984.

Oberg, James E. *Red Star in Orbit.* New York: Random House, 1981.

O'Keefe, Bernard J. *Nuclear Hostages.* Boston: Houghton Mifflin, 1983.

Payne, Keith B., editor. *Laser Weapons in Space: Policy and Doctrine.* Boulder: Westview Press, 1983.

Prados, John. *The Soviet Estimate.* New York: The Dial Press, 1982.

Ritchie, David. *Space War.* New York: Atheneum, 1982.

Royal, Denise. *The Story of J. Robert Oppenheimer.* New York: St. Martin's Press, 1969.

Scheer, Robert. *With Enough Shovels.* New York: Random House, 1982.

Schell, Jonathan. *The Fate of the Earth.* New York: Knopf, 1982.

Stern, Philip M., with the collaboration of Harold P. Green, *The Oppenheimer Case: Security on Trial.* New York: Harper & Row, 1969.

Stine, G. Harry. *Space Power.* Ace Books, 1981.

———. *Confrontation in Space.* Englewood Cliffs, N.J.: Prentice-Hall, Inc., 1981.

Stoiko, Michael. *Soviet Rocketry: Past, Present and Future.* New York: Holt, Rinehart & Winston, 1970.

Taylor, John W. R., editor. *Jane's All the World's Aircraft, 1981–82.* London: Jane's, 1981.

Tsipis, Kosta. *Arsenal: Understanding Weapons in the Nuclear Age.* New York: Simon & Schuster, 1984.

Yergin, Daniel. *Shattered Peace.* Boston: Houghton Mifflin, 1977.